Lab Manual

to accompany

Introductory Horticulture

5th Edition

by

Carroll Shry

Delmar Publishers

I(T)P™ An International Thomson Publishing Company

Albany • Bonn • Boston • Cincinnati • Detroit • London • Madrid • Melbourne
Mexico City • New York • Pacific Grove • Paris • San Francisco • Singapore
Tokyo • Toronto • Washington

Delmar Publishers' Online Services

To access Delmar on the World Wide Web, point your browser to:

http://www.delmar.com/delmar.html

To access through Gopher: gopher://gopher.delmar.com
(Delmar Online is part of "thomson.com," an Internet site
with information on more than 30 publishers
of the International Thomson Publishing organization.)

For information on our products and services:
email: info@delmar.com
or call:
800-347-7707

COPYRIGHT © 1997
By Delmar Publishers
a division of International Thomson Publishing Inc.
The ITP logo is a trademark under license.

Printed in the United States of America.

For more information, contact:

Delmar Publishers
3 Columbia Circle, Box 15015
Albany, New York 12212-5015

International Thomson Publishing Europe
Berkshire House 168 - 173
High Holborn
London WC1V7AA
England

Thomas Nelson Australia
102 Dodds Street
South Melbourne, 3205
Victoria, Australia

Nelson Canada
1120 Birchmount Road
Scarborough, Ontario
Canada M1K5G4

International Thomson Editores
Campos Eliseos 385, Piso 7
Col Polanco
11560 Mexico D F Mexico

International Thomson Publishing Gmbh
Konigswinterer Strasse 418
53227 Bonn
Germany

International Thomson Publishing Asia
221 Henderson Road #05 - 10
Henderson Building
Singapore 0315

International Thomson Publishing - Japan
Hirakawacho Kyowa Building, 3F
2-2-1 Hirakawacho
Chiyoda-ku, 102 Tokyo
Japan

6 7 8 9 10 XXX 02 01 00 99

ISBN: 0-8273-6915-8

Library of Congress Catalog Card Number: 95-15151
CIP

Contents

Preface

This manual has been designed to enhance the lab activities of the *Introductory Horticulture*, Fifth Edition, text. The activities in this lab manual have been used in a technical school setting, which provides the students with necessary practical application skills in math, agriscience, measurement, communication, problem solving, as well as academic background for entry level positions in the horticulture industry and for advanced studies at post-secondary college.

This manual covers the following areas: career exploration; plant nomenclature; plant parts; soils and media for plant growth; basic landscape design; landscape planning; landscape maintenance; power and hand tools of the horticulture industry; and designing holiday arrangements.

The activities in this manual vary in application, from group and crew to individual, using the hands-on approach to learning about the horticulture industry. This lab manual offers a myriad of ways for all students to explore agricultural education.

BARCODES

Barcodes correlating to Delmar Publisher's Agriscience Laserdisc Package are located in all new lab manuals, giving students a visual reinforcement of concepts presented through lab activities. Not all lab exercises will contain a barcode due to the specialized content they cover.

DELMAR PUBLISHERS' AGRISCIENCE LASERDISC PACKAGE

The first laserdisc series developed exclusively to meet the needs of Agriscience programs, the four double-sided discs cover the most widely taught content areas of Agriscience, and correlate directly to fifteen of Delmar's new and best-selling titles. Interactive barcodes for the texts and lab manuals give students instant access to exciting video segments that reinforce concepts presented in class and promote critical thinking through open-ended discussion questions.

Laserdisc Series Package (Discs 1–4)	Order # 0-8273-7300-7
Disc 1: Animal Science	Order # 0-8273-7301-5
Disc 2: Plant Science	Order # 0-8273-7302-3
Disc 3: Business and Mechanical Technology	Order # 0-8273-7303-1
Disc 4: Forestry and Natural Resources Management	Order # 0-8273-7304-X
Pioneer CLD-V2400 Laserdisc	Order # 9999123456
Pioneer Laser Barcode Reader	Order # 9999123457

A Correlation Guide (automatically included with each disc) covering fifteen of Delmar's new and best-selling titles contains barcode images and descriptive captions for each video segment and provides instruction on using the discs effectively with almost any Delmar text.

The Correlation Guide covers:

Volume I: New Products
- *Agricultural Mechanics: Fundamentals and Applications, 3E*
- *Agriscience: Fundamentals and Applications, 2E*
- *Exploring Agriscience*
- *The Science of Agriculture: A Biological Approach*
- *Introductory Horticulture, 5E*
- *Leadership: Personal Development and Career Success*
- *Managing Our Natural Resources, 3E*

Volume II: Best Selling Titles
- *Agriscience and Technology*
- *Aquaculture Science*
- *Ecology of Fish and Wildlife*
- *Environmental Science*
- *Floriculture*
- *Ornamental Horticulture*
- *The Science of Animal Agriculture*
- *Small Animal Care & Management*

Acknowledgments

This manual was developed with the assistance and cooperation of several individuals and organizations. The author and publishers would like to express sincere appreciation to the following:

H. Edward Reiley, for his technical review and suggestions for this lab manual.
Carolina Biological Supply, Burlington, NC, for the Venus Flytrap Tissue Culture Lab 3-13.
Lesco, Inc., Riverbend, OH, for the lab information for Lab 8-7, 8-8, 8-10, 8-11, 8-12.

Thanks are extended to the following for their special assistance:

Wetzel Seed Company, Harrisonburg, VA
Landscape Contractors Association, Rockville, MD
Dutch Plant Farm, Frederick, MD
Treeland Nursery, Inc., Frederick, MD
Stadler Nursery, Onley and Frederick, MD
Chapel Valley Landscape, Inc., Lisbon, MD
Bluemont Nursery, Monkton, MD
Frederick County Career and Technology Center, Frederick, MD

Judy Shry for her continued support and encouragement during this project.

SECTION 1

Horticulture:
An Introduction

◆

LAB EXERCISE 1–1

Exploring Employment Opportunities in Horticulture

PURPOSE
To explore the employment opportunities in the horticulture field

MATERIALS
pen or pencil
Introductory Horticulture, 5th Edition

PROCEDURE

1. Using the information given in the job analysis chart provided on the next page, select five possible careers that interest you and give eight interesting facts about each career.

_____

_____

_____

_____

_____

_____

_____

_____

2. Using the local newspaper, search the employment section for job opportunities in your area.

3. Using the local telephone directory yellow pages, find a listing of the local horticulture businesses. Make a list of ten business names and phone numbers.

_____

_____

_____

_____

_____

_____

_____

_____

_____

4. Write to your state horticultural society or state nurserymen's association and ask the group's president to speak to the class about job opportunities.

5. Read an article in a horticulture trade magazine (e.g., *American Nurserymen, Florist Review, Turf, Landscape Architecture*) and write a summary of the article and present an oral report to the class on your findings.

6. Visit a local garden center owner/operator to explore job opportunities.

JOB ANALYSIS CHART

JOB TITLE	Does it entail year-round work?	Are there regular hours?	Is most of the work outdoors or indoors?	Does it offer variety?	Is the work in one place?	Are there fringe benefits?	Does the job involve working with others?	What are the educational requirements?	Is there an opportunity for promotion?
greenhouse worker	yes	Generally, but some overtime is usually required.	indoors	yes	yes	yes	Yes, to some extent.	high school diploma with a course in agriculture or horticulture	yes
nursery worker	in many cases	Yes, but there are peak seasons.	mostly outdoors	yes	yes	some	yes	high school diploma with a course in agriculture or horticulture	yes
garden center employee	yes	Yes, but there are peak seasons.	both indoors and outdoors	yes	yes	some	yes	high school diploma with a course in agriculture or horticulture	yes
golf course employee	no	Yes, during the golfing season.	outdoors	yes	yes	not to a great extent	yes	high school diploma with a course in agriculture or horticulture	yes
assistant grounds keeper	yes	Yes, some overtime is required on occasion.	mostly outdoors	yes	yes	There may be some.	not necessarily	high school diploma with a course in agriculture or horticulture	yes
park employee	yes	yes	outdoors	yes	yes	not to a great extent	yes	high school diploma with a course in agriculture or horticulture	yes
vegetable grower	depends upon grower	no, seasonal	outdoors	yes	yes	some	yes	high school diploma with a course in agriculture or horticulture	yes
orchard employee	depends upon grower	no, seasonal	outdoors	yes	yes	some	yes	high school diploma with a course in agriculture or horticulture	yes
employee of small fruit grower	depends upon grower	no, seasonal	outdoors	yes	yes	some	yes	high school diploma with a course in agriculture or horticulture	yes
employee of floral design shop	yes	yes, with some overtime	indoors	no	yes	yes	yes	high school diploma with a course in agriculture or horticulture	yes

LAB EXERCISE 1–2

Exploring the Horticulture Field

PURPOSE
To explore the myriad career choices in the horticulture field

MATERIALS
pen or pencil
Introductory Horticulture, 5th Edition

PROCEDURE
Read Unit 1 and answer the following questions.

1. Define horticulture. _____

2. What are the four major divisions of the horticulture industry?

    (1) _____

    (2) _____

    (3) _____

    (4) _____

3. List four basic businesses that employ individuals who are training in horticulture.

    (1) _____

    (2) _____

    (3) _____

    (4) _____

4. Describe the working conditions of a greenhouse employee. _____

    _____

5. If you enjoy working outdoors, a job in a nursery might be for you. List the qualifications for nursery employees. _____

    _____

6. Describe the working conditions of a garden center employee. _____

    _____

7. What kind of work does a ground maintenance employee do? _____

    _____

8. If you worked as a golf course employee, what would your responsibilities be? _____

    _____

9. What is the job title of an employee who maintains a park? _____

    _____

10. List the opportunities a college graduate with a degree in horticulture would have. _____

    _____

11. The horticulture industry has opportunities in the scientific field and the sales and service area.
    List three jobs in each of these areas.

    (1)_____

    (2)_____

    (3)_____

LAB EXERCISE 1–3

Horticulture Crossword Puzzle

PURPOSE

To complete the crossword puzzle using your knowledge of horticulture

MATERIALS

pen or pencil

PROCEDURE

Fill in the crossword puzzle on the next page using the words below to answer the "Down" clues and the "Across" clues.

Parks	Skills	Landscape	Conservation
Pests	Sod	Olericulture	Duties
Pomology	Green	Ornamental	Environment
Prune	Greenhouse	Orchards	Floriculture
Seedling	Horticulture	Aerate	Florist
Seed	Graft	Career	Golf

DOWN CLUES

1. The science of growing and getting a crop of fruit to the marketplace.
2. Someone who grows, arranges, and sells flowers.
4. The science of growing vegetables for profit.
5. A very young plant.
7. A place designed to grow plants indoors.
8. Places that require the upkeep of the grass, trees, and shrubs to maintain a recreational area.
10. Places that grow apples, peaches, pears, and so forth.
11. The thing that a plant makes to reproduce itself.
12. The caring for a garden.
14. Grass and its roots with a thin layer of dirt.
17. Insects, moes, mice, rabbits, and humans.
19. To loosen soil so that air can get below the surface.
20. The sport that provides jobs for landscapers.

ACROSS CLUES

3. The science of profitably making flowers available.
6. Our _____ is everything that is and/or around where we live.
9. To keep from losing what we have been given or have worked for.
13. Those things that we learn to do well.
15. The "putting" area of a golf course.
16. Certain jobs and tasks assigned only to responsible persons.
17. To cut back the branches of a plant to control growth.
18. A job we enjoy doing and are willing to learn how to do right so someone will hire us.
21. The cutting and joining together of two separate plants.
22. Something whose main purpose is to add beauty.
23. What we do to land and the plants to make it attractive and serve our purpose.

NAME: _____ DATE: _____

LAB EXERCISE 1–4

Personal Data Sheet or Resume

PURPOSE

To prepare a Personal Data Sheet or Resume

MATERIALS

pen or pencil

PROCEDURE

Using the sample resume provided below as an example, prepare your own resume by filling in the spaces in the following blank resume.

RESUME

Steve C. Cover

468 Hopemont Street
Woodsboro, Maryland 21798
301-663-8007

Career Objective: Seek a position with a commercial landscaping company

Education: Senior at Delmar High School, to graduate in May, 1997 .
Majoring in Landscape and Nursery Management

Subjects Studied:
Algebra:	1 semester
AgriScience:	2 semesters
Biology:	2 semesters
Chemistry:	1 semester
Consumer Economics:	1 semester
Cooperative Work Experience:	2 semesters
English:	4 semesters
Landscape and Nursery Production:	2 semesters

Student Activities: President of the FFA Chapter
Treasurer of the Senior Class
Member of the School Band

Work Experience: September 1995 to present
Blue Mountain Landscpe Company
133 Dogwood Lane, Frederick, Maryland 21701

References: 1. Mr. Ron Smith
7999 River Road, Walkersville, Maryland 21793

2. Mrs. Linda Summers
243 Nolands Ferry Road, Tuscarora, Maryland 21701

NAME: _____ DATE: _____

RESUME

Name: _____

Address: _____

Telephone: _____

Career Objective: _____

Education: _____

_____

Subjects Studied: _____

_____

_____

_____

_____

_____

Student Activities: _____

_____

_____

_____

Work Experience: _____

_____

_____

References: _____

_____

_____

NAME: _____   DATE: _____

LAB EXERCISE 1–5

Job Application

PURPOSE

To complete a job application

MATERIALS

pen or pencil

PROCEDURE

Using the information from the resume you completed in Lab Exercise 1–4, complete the job application on the following pages.

APPLICATION FOR EMPLOYMENT

All applicants will be considered for employment without regard to race, religion, color, sex, national origin, age, marital status, disability, sexual orientation, veteran status or any other status protected by law. We are an Equal Opportunity Employer.

PERSONAL INFORMATION
(please print)   Date _____

Name _____ Soc. Sec.# _____

Address _____ City _____ State _____ Zip _____

Telephone No. _____ Referred By: _____

Are you legally eligible for permanent employment in the U.S.? _____

Position(s) applied for _____ Full Time ☐   Part Time ☐

Seasonal Help only ☐ Yes ☐ No

If part time, specify days/hours available _____

Rate of pay desired: $ _____ per _____

Have you worked for us before? _____ If YES, when? _____ Position _____

Indicate special qualifications or skills _____

_____

EDUCATION NAME & LOCATION OF SCHOOL	COURSE OF STUDY	DID YOU GRADUATE?	IF YOU DIDN'T GRADUATE YEARS COMPLETED OR ATTENDED
ELEMENTARY			
HIGH SCHOOL			
COLLEGE	MAJOR		
	DEGREE		
OTHER			

SPECIAL QUESTIONS

Are you over 18 years of age? _____ If NO, state your age: _____ (employment subject to legal age verification)

** IT IS UNLAWFUL IN MA & MD TO REQUIRE OR ADMINISTER A LIE DETECTOR TEST AS A CONDITION OF EMPLOYMENT OR CONTINUED EMPLOYMENT. AN EMPLOYER WHO VIOLATES THIS LAW SHALL BE SUBJECT TO CRIMINAL PENALTIES AND CIVIL LIABILITY.

(CONTINUED ON OTHER SIDE)

NAME: _____ DATE: _____

PRIOR EMPLOYMENT *(Start with most recent employer; you may also include volunteer work)*

Employer:	Phone:	From:	To:
Address:	City, State, Zip	Position:	
Duties:		Supervisor's Name:	
		Starting Salary/Wages:	
Reason for leaving:		Final Salary/Wages:	
Employer:	Phone:	From:	To:
Address:	City, State, Zip	Position:	
Duties:		Supervisor's Name:	
		Starting Salary/Wages:	
Reason for leaving:		Final Salary/Wages:	
Employer:	Phone:	From:	To:
Address:	City, State, Zip	Position:	
Duties:		Supervisor's Name:	
		Starting Salary/Wages:	
Reason for leaving:		Final Salary/Wages:	

MILITARY SERVICE

BRANCH OF SERVICE	FROM	TO	RANK & DUTIES	DATE DISCHARGED

BUSINESS REFERENCES *(Please supply us with three business references, preferably current or former supervisors. If unable to supply three supervisors please give coworkers or subordinates.)*

NAME	ADDRESS	YEARS KNOWN	TELEPHONE

I authorize investigation of all statements contained in this application. I understand that misrepresentation, falsification or omission of facts called for is cause for dismissal. Further, I understand and agree that my employment is for no definite period and may be terminated at any time without any previous notice.

Signature: _____

REMARKS:

LAB EXERCISE 1–6

Agriscience Careers Word Search

PURPOSE

To develop a vocabulary of horticulture careers by completing the word search

MATERIALS

pen or pencil

PROCEDURE

Find these words in the word search below.

horticulturist	agronomist	hydroponist	greenskeeper
floriculturist	florist	geneticist	limnologist
embryologist	microbiologist	zoologist	taxonomist
bioengineer	viticulturist	bacteriologist	forester
pomologist	ichthyologist	entomologist	

```
t o k n d v z k l d g e w i u f t m m r r h p
a m e x j s o q p i n o c r l g i x u e p o s
x k w h r h k p m t e h c o o c t x e p x r r
o h a m q p l t o o t l r j r c s n e e v t o
n w y n i x l m u h p i e o b l i s i e i i o
o h n d r y o d y t c t b b w g g a k k t c m
m r z o r l x o t u s i m a n k o f h s i u w
i m p a o o l t l i o f f e g t l c a n c l r
s q m g m o p t g l h v o p m r o c m e u t a
t m i b g w u o o x k i g r h d o r t e l u c
f s j i v r l g n d b q y m e s z n n r t r m
t l s v i o i a o i m x y b l s l q o g u i j
z t o s y s o a w i s o g p m l t l x m r s s
r z t r t r d g u z z t t h w z o e u i i t x
y b b u i i q f i g h m n v x o f n r o s s z
c m c a t s i g o l o i r e t c a b x t t e t
e e j p p o t r t n d t s i g o l o n m i l f
t s i g o l o m o p e h b d y t b k e k y z v
s a s o g e n e t i c i s t h b r x y v e c g
```

LAB EXERCISE 1–7

Bulletin Board on Careers in Landscaping

PURPOSE

To develop a bulletin board about careers in landscaping

MATERIALS

current trade magazines in the following topic areas: landscape maintenance, grounds and golf course maintenance, such as *American Nurseryman Magazine*, *Turf*, *Florist Review*, *American Horticulturist*, *Grounds Maintenance*, *Pro*, *Flower and Garden*, *Landscape Design*, *Landscape Architecture*, *Horticulture*, *Garden Designs*.
local newspaper employment section
pen or pencil

PROCEDURE

After completing the following activities, develop a bulletin board incorporating the information, articles, job notices, etc., you gathered.

Activities

1. Review current trade magazines on landscape maintenance and grounds and golf course maintenance, such as *American Nurseryman Magazine*.

2. Review textbook reference on Careers in Landscape.

3. Read local newspaper for available jobs in the landscaping business.

4. Discuss career opportunities in landscaping with your instructor.

LAB EXERCISE 1–8

Plant Nomenclature Fill-in-the-Blank

PURPOSE

To complete the fill-in-the-blank questions using your knowledge of nomenclature

Plants have scientific and common names; it is important to understand the use of both names. In the horticultural industry the scientific name is standard, while the common name is used by laymen.

MATERIALS

pen or pencil
Introductory Horticulture, 5th Edition

PROCEDURE

Read pages 13 to 16 in Unit 2 and answer the following questions.

1. The Redbud tree and _____ are the same tree.

2. Give an example of one common name that represents two different plants: _____ .

3. The Swedish botanist Linnaeus classified plants into the _____ system.

4. All scientific names of plants are expressed in _____ .

5. The generic name is a _____ and the species name is an adjective.

6. When a plant name lists the genus and species with 'cv,' this means the plant is a _____

   _____

7. When Latin names are printed they are expressed in _____ .

8. Using the binomial system, what name is written first? _____

9. Scientists that classify plants are known as _____ .

10. The first letter of the genus in the binomial system is always _____ , while the first letter in the species is always in the _____ case.

LAB EXERCISE 1–9

Plant Science
Side 1
Classifying Plants

Plant Taxonomy Word Match

PURPOSE

To complete the word match using your knowledge of plant taxonomy

The plants' scientific names are expressed in Latin because Latin is a universal language. The scientific names will be in italics.

MATERIALS

pen or pencil

PROCEDURE

Match the common name with the genus/species.

1. _____ Birch
2. _____ Red Oak
3. _____ Tobacco
4. _____ Yew
5. _____ Catnip
6. _____ Geranium
7. _____ Trailing arbutus
8. _____ Pinks
9. _____ Pine
10. _____ Yellow Birch
11. _____ White Pine
12. _____ Walnut
13. _____ Red Maple
14. _____ Butterfly bush
15. _____ Sunflower
16. _____ Peach
17. _____ Mum
18. _____ Maple
19. _____ Poppy
20. _____ Oak

A. *Epigaea repens*
B. *Acer*
C. *Acer rubrum*
D. *Quercus boralis*
E. *Dianthus*
F. *Betula*
G. *Papaver*
H. *Buddleia davidi*
I. *Pinus strobus*
J. *Nicotinia*
K. *Prunus persica*
L. *Nepeta*
M. *Quercus*
N. *Pinus*
O. *Taxus*
P. *Chrysanthemum*
Q. *Juglans*
R. *Pelargonium*
S. *Betula lutea*
T. *Helianthus*

LAB EXERCISE 1–10

Plant Science
Side 1
Plant Anatomy & Physiology

Plant Taxonomy: Using the Flower for Identification

PURPOSE

To learn to recognize different plants by using the morphology of the different flower structures

A scientist who identifies and classifies plants is known as a taxonomist. This skill will prepare you for many horticultural opportunities.

MATERIALS

pen or pencil

PROCEDURE

Using the figure below as a guide, draw and label the following flowers.

1. Simple (complete) Flower (e.g., daffodil)

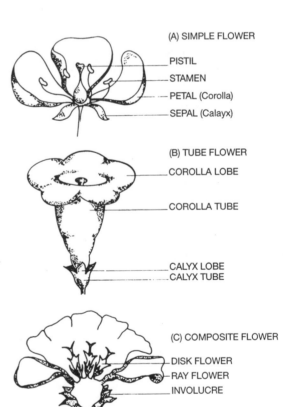

(A) SIMPLE FLOWER
PISTIL
STAMEN
PETAL (Corolla)
SEPAL (Calayx)

(B) TUBE FLOWER
COROLLA LOBE
COROLLA TUBE
CALYX LOBE
CALYX TUBE

(C) COMPOSITE FLOWER
DISK FLOWER
RAY FLOWER
INVOLUCRE

2. Tube Flower (e.g., petunia or foxglove)

3. Composite Flower (e.g., daisy, chrysanthemum)

4. Incomplete Flower (e.g., male or female holly)
   Note: this flower will be missing the pistil or stamen.

SECTION 2

Plant Science

◆

NAME: _____   DATE: _____

LAB EXERCISE 2–1

Leaf Collection

PURPOSE

To make a leaf collection of deciduous trees to help identify trees used in the horticultural industry by recognizing the leaf form and margins

MATERIALS

leaf collection press
3″ x 5″ file cards for labels
Introductory Horticulture, 5th Edition
hand pruners

PROCEDURE

1. Make a leaf collection of ten deciduous trees in your area. (Using your school grounds or local park to collect samples of deciduous trees will help to identify the trees and recognize the different leaf forms and margins.

2. When you return from your leaf collection expedition, your leaves need to be placed in a leaf press to preserve the leaf quality for future reference.

   a) Place the leaf between absorbent leaf press sheets.

   b) Using *Introductory Horticulture*, 5th Edition, identify the different leaf forms and margins.

   c) Label each leaf with its name, location of collection, and date. (Several leaves may be placed on one sheet with its label when using the leaf press.)

   d) After four to six weeks the leaves may be removed from the press and mounted in your taxonomy notebook for class.

3. Using one of the leaves you collected, label the major parts of the leaf including: blade, petiole, margin, tip, midrib, veins, and base.

   a) Tape a leaf to a sheet of paper.

   b) Using, *Introductory Horticulture,* 5th Edition, label the major parts.

LAB EXERCISE 2–2

Plant Science
Side 1
Plant Anatomy & Physiology

Plant Parts and Functions Fill-in-the-Blank

PURPOSE

To fill in the blanks with the correct answers using your knowledge of plant parts and functions

MATERIALS

pen or pencil
Introductory Horticulture, 5th Edition

PROCEDURE

Answer the following questions.

1. Plants are the primary source of _____ for humans and animals.

2. What is an essential product needed by all living things that is produced by plants?

   _____

3. Name five things that plants supply us.

   _____

   _____

   _____

   _____

   _____

4. What is the major function of the flower of a plant? _____

5. In addition to providing absorption of water for the plant, what other functions do the roots provide? _____

6. What part of the plant is known as the food factory? _____

7. What are three ways leaves are arranged on the stem? _____

8. What cells allow the plant to transpire water? _____

9. What cells give the plant leaf its color? _____

10. Write the chemist shorthand equation for photosynthesis. _____

11. What must be present for photosynthesis to take place in a plant? _____

12. The oxygen produced by the photosynthesis process is a vital ingredient in _____

_____

13. What do plants consume during the respiration and growth processes? _____

14. List two functions of the stem.

_____

_____

15. _____ is the name of the breathing pores on the stem.

16. Collect deciduous plant stem samples and label the major parts of the stem.

17. The stem of woody deciduous plants is composed of bark called _____ and wood called _____.

18. List two major differences between the monocots and dicots.

_____

_____

19. What are the functions of xylem and phloem? _____

20. What absorbs water and minerals on the roots? _____

LAB EXERCISE 2–3

Plant Science
Side 1
Plant Anatomy & Physiology

Plant and Flower Part Identification

PURPOSE

To identify the parts of plants and flowers

MATERIALS

pen or pencil
flower (e.g., lily, daffodil, amaryllis, or tulip)
scalpel
tape

CAUTION
Scalpels are sharp!

PROCEDURE

1. Select a flower.
2. Using the scalpel dissect the flower parts.
3. Identify the number of parts and attach the parts to a 3″ × 5″ card.
4. Label the parts of a complete flower: Use the diagram below as a guide.

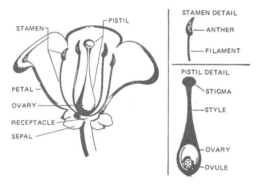

Complete the following questions.

1. Insects, wind, and rain aid in the transfer of pollen from the stamen to the pistil called _____
   _____ .

2. Seeds are produced by a _____ with a male and female parent involved.

3. A _____ has both male and female parts.

4. When the pistil is receptive, the _____ will be sticky to catch the pollen grains.

5. The incomplete flower is missing _____ .

6. What part of the flower is the male reproductive part? _____

7. After fertilization, the _____ turns into a fruit or seed pod.

LAB EXERCISE 2–4

Horticulture Crossword Puzzle

PURPOSE

To complete the crossword puzzle using your knowledge of horticulture

MATERIALS

pen or pencil

PROCEDURE

Fill in the crossword puzzle on the next page using the words below to answer the "Down" clues and the "Across" clues.

Ace	Annual	Arborists	Chlorotic	Deciduous
Doggy	Dripline	Emulsion	Foliage	Foliat
Phosphorus	Prune	Punch	Stub	Succulent
Sucker	Systemic	Taxus	Topiary	Toxic
Girdling	Herbicide	Heading	Mole	Mold
Nursery	Nutrient	Peat	Perennial	Pesticide

DOWN CLUES

1 Selectively trim a plant.
2 A selective plant poison.
5 A plant that lives for only one year.
9 What need to be trimmed flush.
10 People who work with trees.
12 Decayed plant matter.
13 A type of mixture.
15 A furry growth.
17 A type of pruning.
18 Growth beneath a graft.
19 The outer watershed of a tree.
20 A plant that sheds its leaves once a year.
24 Having to do with leaves.
25 _____bar fertilization.
26 Something harmful or poisonous.
27 A bag you get in a reastaurant.

ACROSS CLUES

3 A place of plant nurturing.
4 Scrub sculpture.
6 A plant with a life greater than two years.
7 Highest card in the deck.
8 Having to do with too much growth.
11 Something needed to sustain life.
14 leaves
16 One of the basic nutrients.
21 Something that affects a whole organism.
22 A little blind mammal.
23 Loss of color.
25 Something to kill the unwanted.
28 Juicy.
29 Surrounding.

LAB EXERCISE 2–5

Plant Parts and Their Function Word Match

PURPOSE

To accurately match the plant parts with their functions

MATERIALS

pen or pencil
Introductory Horticulture, 5th Edition

PROCEDURE

Match the descriptions with the plant parts and/or functions of plants.

1. _____ Flower
2. _____ Leaves
3. _____ Fruit
4. _____ Seed
5. _____ Stem
6. _____ Roots
7. _____ Epidermis
8. _____ Guard cells
9. _____ Stoma
10. _____ Transpire
11. _____ Chloroplasts
12. _____ Photosynthesis
13. _____ Glucose
14. _____ Kilocalorie
15. _____ Oxidation
16. _____ Petiole
17. _____ Cuticle
18. _____ Midrib
19. _____ Blade
20. _____ Margin

A. Skin of the leaf
B. Food manufacture by photosynthesis
C. Seed carrying structure
D. Sugar produced during photosynthesis
E. Attracts insects for pollination
F. Process of manufacturing food
G. Green color in the leaf
H. Main support of the plant passage for food and water
I. Process of combining oxygen (e.g., rotting)
J. Exchange oxygen and carbon dioxide
K. Sexual propagation of plants
L. Measurement of light energy
M. Anchors the plant, absorbs food, storage
N. Special cells in the epidermis
O. Small pore that allows the leaf to breathe
P. Edges of the plant leaf
Q. Larger center portion of the leaf from other leaf veins extend
R. Leaf stalk
S. Large flat part of the leaf
T. Layer covering the epidermis

LAB EXERCISE 2–6

Plant Parts Vocabulary Word Match

PURPOSE

To complete the word match using your knowledge of plant parts

MATERIALS

pen or pencil
Introductory Horticulture, 5th Edition

PROCEDURE

Match the description with the plant parts.

1. _____	Terminal bud	A.	Moves plant food to the roots
2. _____	Axillary bud	B.	Breathing pores on the stem
3. _____	Node	C.	Moves water and nutrients to leaves and stem
4. _____	Internode	D.	Layer separating xylem and phloem
5. _____	Abscission layer	E.	Plants that produce seed leaves called cotyledons
6. _____	Leaf scar	F.	Center of the stem
7. _____	Bud scales	G.	Indicate where the terminal bud was located previous year
8. _____	Lenticel	H.	Leaf axis from which flowers and branches rise
9. _____	Dicots	I.	Absorbs moisture and minerals
10. _____	Monocots	J.	Point where the petiole attached the stem
11. _____	Cambium	K.	Part of the stem between two nodes
12. _____	Xylem	L.	Plants that are grasses
13. _____	Phloem	M.	Tip of shoot where growth takes place
14. _____	Pith	N.	The cell layer where the branches and leaves separate
15. _____	Root hairs	O.	Where the branch attaches the stem
16. _____	Alternate	P.	Leaf has the shape of the palm of the hand (e.g., maple)
17. _____	Opposite	Q.	Leaves encircle the petiole of the plant (e.g., Japanese andromedia)
18. _____	Whorled	R.	Leaf arrangement of the flowering cherry which are off-set from each other
19. _____	Pinnate, Compound	S.	Smaller individual leaflets make up the complete leaf
20. _____	Palmate	T.	Leaf arrangement of the Japanese maple (leaves are across from each other)

LAB EXERCISE 2–7

Plant Science
Side 1
Plant Anatomy & Physiology

External Stem Structure

PURPOSE

To recognize the external stem structure of the plant, which will help each student identify landscape plant material used in today's landscapes

MATERIALS

3″ × 5″ index cards
live tree samples
Introductory Horticulture, 5th Edition
hand pruner

⚠ **CAUTION**
Use care when handling
sharp hand pruners.

PROCEDURE

In this lab exercise you will collect five samples of stems of deciduous trees or shrubs to study the exterior parts of the plant. In the process, please note some deciduous plants have opposite leaf arrangements, while other genera have the alternate leaf arrangement. An easy way to remember the plants with opposite leaf arrangements is MAD HORSE (maple, ash, dogwood, and horse chestnut).

The outside stem consists of a terminal bud, which is the tip or end shoot of the stem where growth takes place. The node is where the leaf petiole is attached to the stem. The distance between two nodes is called the internode. The axillary bud is located on the side of the stem in the leaf axis from which flowers and branches grow. Located between the leaf and the branch is a thin cell layer that separates in the fall on deciduous plants called the abscission layer. Also on the stem are lenticels which allow the stem to breathe.

Label the major exterior parts of the plant on the figure below.

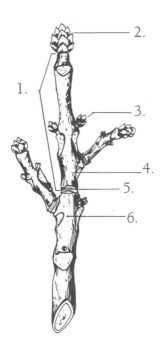

LAB EXERCISE 2–8

How Plants Grow Word Match

PURPOSE

To complete the word match using your knowledge of how plants grow

MATERIALS

pen or pencil
Introductory Horticulture, 5th Edition

PROCEDURE

Match the term with its definition.

1. _____ Cambium

2. _____ Phloem

3. _____ Seed

4. _____ Sepal, Petal, Pistil

5. _____ Leaf, Stem, Root

6. _____ Irish potato

7. _____ Oxygen/Food

8. _____ Fibrous

9. _____ Root hairs

10. _____ Xylem

A. Water and minerals travel up the stem

B. Part of the stem found in dicots which allow the plant to produce new cells

C. Absorbs water and minerals from the soil

D. Main parts of the flower

E. Basic parts of the plant

F. Most important thing(s) humans receive from plants

G. Plants with root systems that are easier to transplant than plants with a tap root

H. Underground stem that stores food

I. Manufactured food travels down the stem

J. Flower produces this structure when it is fertilized

LAB EXERCISE 2–9

Plant Science
Side 2
Tree Fruits and Nuts

Cross Section of a Stem

PURPOSE

To draw a cross section of a stem to identify the various parts of the stem, which will help you prune plants in the future

MATERIALS

stereomicroscope
hand pruners
pruning saw
lopping shears
Introductory Horticulture, 5th Edition

PROCEDURE

In this lab exercise you will collect different samples of tree branches using hand pruners, lopping shears, or pruning saws, to obtain a representative sample of various genera of deciduous trees. Some suggested trees are: tree of heaven, linden, maple, beech, tulip poplar, and birch.

Answer the following questions.

1. Name the two main functions of stems.

   (1)_____

   (2)_____

2. What are the breathing pores on the stem called? _____

3. Using one of the stem samples you collected, attach it to a 3″ × 5″ card and label the parts of the stem. (Note: in some cases you may have to use a stereomicroscope to determine the primary and secondary phloem.)

4. Make small cross sections of the deciduous tree samples and tape them to a sheet of paper. Label the parts.

5. The stem of woody plants is composed of bark called _____ and wood called _____.

6. What is the major difference between monocots and dicots? _____

   _____

7. What does phloem and xylem allow the plant to do? _____

   _____

8. What on the roots absorbs water and minerals? _____

9. Name two major root systems.

(1)_____

(2)_____

10. Which of the major root systems is easier to transplant? _____

LAB EXERCISE 2–10

Collecting and Preparing Soil Samples

PURPOSE

To collect and prepare soil samples for testing, so as to obtain the most accurate soil sample representation of the sampled area

MATERIALS

two plastic pails (one to mix the composite sample)
plastic storage bag or bucket
soil probe/soil auger
nursery spade
Introductory Horticulture, 5th Edition

PROCEDURE

When collecting soil samples, it is important to remember the samples must be representative of the area. In this lab exercise, you will be making a composite sample of approximately a pint of air-dried soil to be tested.

1. Sketch out a map of the area to be tested.

2. Divide the area into manageable sites to test.

3. Collect a composite sample from each of the areas to be tested.

4. Probe into the soil 4 to 6 inches.

5. Take 12 to 15 core samples from each of the areas to be tested. Mix the core samples together to make your composite sample.

6. Do not sample from unusual spots (e.g., low areas).

7. Be sure the containers are properly labeled.

8. Allow collected samples to air dry before testing.

LAB EXERCISE 2–11

Plant Science
Side 2
Irrigation and Crop Protection

Soil Testing

PURPOSE

To learn how to test the soil samples you collected in Lab Exercise 2–10 to make the correct recommendations for optimum plant growth

MATERIALS

soil testing kit
distilled water

PROCEDURE

It is important to analyze the soil samples collected in order to supply the correct nutrients for optimum plant growth. In this exercise you test the soil for its pH level and the macronutrients of nitrogen (N), phosphorous (P), and potassium (K).

1. Read the directions supplied with the Sudbury Soil Testing Kit.

2. Test for pH level and N, P, K macronutrients in each of the soil samples.

3. Use the following table to record your findings.

LABORATORY SOIL SAMPLE INFORMATION

Sample No.	Location in the Field	pH	N	P	K	Recommendations

LAB EXERCISE 2–12

Plant Science
Side 2
Irrigation and Crop Protection

Soil Testing Fill-in-the-Blank

PURPOSE
To understand how to correct the pH by using soil amendments

MATERIALS
pen or pencil
Introductory Horticulture, 5th Edition

PROCEDURE
Read Unit 4 in your text and answer the following questions.

1. What is the ideal pH of most plants? _____

2. A pH of 7 represents a soil that is _____. A pH of 7 is neither
   _____ nor _____ .

3. The pH scale goes from _____ to _____ .

4. Values lower than 7 on the pH scale represent an _____ soil. Values higher than
   7 represent an _____ soil.

5. What determines if a soil is acid or alkaline? _____

6. What parts of the country have acid soil? _____

   Why? _____

7. What parts of the country have alkaline soil? _____

   Why? _____

8. What is used to lower the pH of the soil? _____

9. How do you determine what the pH of the soil is? _____

10. What does lime affect in the soil? _____

11. Why does the release of iron (Fe) and aluminum (Al) create a problem in the soil? _____

12. Draw the pH scale, labeling pH values and where plants grow best. _____

13. What does adding lime to the soil activate? _____

14. The addition of lime activates _____ and encourages the release of
    _____ .

15. How can soil structure be improved? _____
    _____

LAB EXERCISE 2–13

Determining Soil Particle Size

PURPOSE

To determine the different soil particle sizes in a sample of soil, which influence the movement of air and water in the soil

MATERIALS

wide mouth pint jars
soil samples
liquid detergent

PROCEDURE

1. Select a wide mouth canning jar (as shown to the right) and fill half-full with soil. Do not compact the soil in the jar.

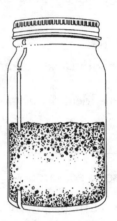

2. Add water until the jar is two-thirds full and add two drops of liquid detergent, which serves as a wetting agent.

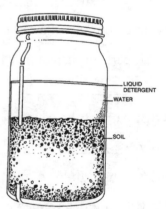

3. Tighten the lid and shake the sample, mixing the soil, water, and wetting agent. Mix the sample thoroughly for 5 minutes. Allow the sample to stand to settle the particles. Observe the sample as it settles at 5 minutes, 10 minutes, 20 minutes, 2 hours, and 24 hours.

   What results did you observe?

   _____

   _____

   _____

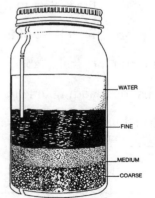

LAB EXERCISE 2–14

Using a Soil Survey Map

PURPOSE

To learn how to interpret a soil survey map for future landscape use

MATERIALS

pen or pencil
soil survey map for the county in which you live

PROCEDURE

This lab uses the soil survey maps developed on a county basis by each state's U. S. Soil Conservation Service. These surveys give an extensive account of the basic data of soils, including:

- A summary of the major soil types on the property and their potential suitability for use;
- An aerial photograph of the area to scale, with the major soils identified and mapped out with boundary lines;
- A summary of possible land uses and the physical limitations on the property.

Activities

1. Using a soil survey map, locate the area where you live.

2. Find the different soils on your property. Write down the code HcD2. Using the code, refer to the index and find the description of the major soil type. Record and write findings on each of the major soil types found in the area.

3. What is the major land use of the area?

_____

_____

_____

_____

LAB EXERCISE 2–15

Determining Soil Texture

PURPOSE

To determine the soil texture using your sense of touch so you can evaluate the quality of soil to produce optimum growth

MATERIALS

clay
flour
sand
soil samples of clay loam and sandy soil
Introductory Horticulture, 5th Edition

PROCEDURE

To determine the texture of a soil by touch, the soil sample is rubbed between your thumb and index finger. To practice what the samples should feel like, use pottery clay and work it between your thumb and index finger. Notice the smooth ribbon it forms. (If the clay is dry, moisten it to give a plastic feel.) This identifies soil as *fine texture*.

Repeat the procedure with flour, which will feel smooth with a slight grit. This identifies soil as *medium texture*.

Repeat the procedure with sand, which will feel gritty (like beach sand). This identifies soil as *coarse texture*.

The texture of the soil is the most fundamental influence of all aspects of the soil's physical characteristics. The soil particle size is the basic factor that determines the texture of the soil. The three basic particles of the soil are: sand (largest); silt (medium); and clay (smallest). These traits have a direct influence on the soil's water-holding capacity, permeability, plant growth, and land classification. The following diagrams show the relative size of the basic soil particles sand, silt, and clay.

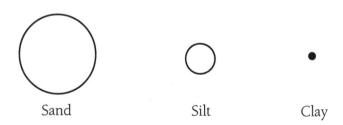

Sand   Silt   Clay

NAME: _____   DATE: _____

Follow these steps to identify basic soil texture by touch.

STEP 1: Start with 30 grams of soil in your hand. Working the soil in the palm of your hand, add water in drops until it forms a putty consistency. (If you make the soil too wet, add more soil.)

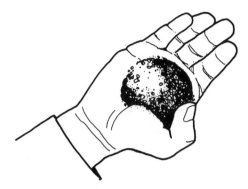

STEP 2: Placing the ball of soil between your thumb and forefinger, work the soil with your thumb to form a ribbon. Allow the soil ribbon to form so it extends over your forefinger and breaks off from its own weight.

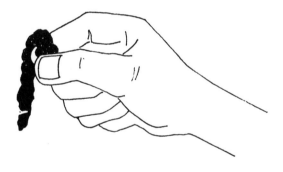

Does this soil form a ribbon?   ❏ Yes   ❏ No

If no, then this means the soil is coarse texture. If yes, then ask yourself the following questions: Does the soil make a ribbon of 1 to 2 inches? If so, then it is medium texture. Does the soil make a ribbon of 2 inches or more? If so, then it is fine texture.

Review the information in Unit 4 of your text to answer the following questions:

1. How are soils classified? _____

_____

2. List the soil particle sizes from the largest to the smallest. _____

_____

3. Draw and label a soil profile.

4. Draw and label the parts of soil composition.

5. Why is air important in the composition of soil? _____

_____

_____

LAB EXERCISE 2–16

Plant Science
Side 1
Soil

Using the Textural Triangle Chart

PURPOSE

To learn how to use the texture triangle

MATERIALS

pen or pencil

PROCEDURE

When the soil laboratory does a mechanical analysis of the soils being evaluated, the process will determine the percentage of sand, silt, and clay in the soil. There are twelve textural classes in the texture triangle. Each side of the triangle is a soil particle percentage ranging from 0 percent to 100 percent. Each of the corners of the triangle is representative of sand, silt, and clay.

Loam soil is an equal mixture of sand, silt, and clay. It is considered to be the ideal soil type for many horticulture crops, while soils with a higher sand content are considered to be coarse with good to excessive drainage, low cation exchange capacity, and poor water-holding capacity. Clay particles, the smallest particles of the three, have the greatest cation exchange capacity.

The diagram below is a guide for textural classification. Use it to determine the textures of the soils that follow.

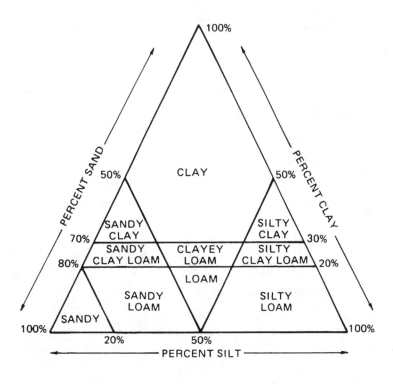

1. Name the texture for each of the following soils.

   A. 50% clay, 30% silt, 20% sand _____

   B. 50% clay, 50% sand _____

   C. 12% clay, 10% silt, 78% sand _____

   D. 22% clay, 60% silt, 18% sand _____

   E. 30% clay, 40% silt, 30% sand _____

   F. 35% clay, 42% silt, 23% sand _____

   G. 30% clay, 60% silt, 10% sand _____

   H. 50% clay, 45% silt, 5% sand _____

   I. 40% clay, 10% silt, 50% sand _____

2. Using the diagram on the previous page, determine what is the minimum percentage of clay a soil may contain and still be considered to be a clay texture? _____

   a loam texture? _____

   a sand texture? _____

LAB EXERCISE 2–17

Plant Science
Side 2
Irrigation and Crop Protection

Controlling Soil Erosion

PURPOSE

To compare the different amounts and rates of soil erosion on land applications

Soil erosion causes the loss of valuable top soil used for plant growth.

MATERIALS

3 greenhouse flats
400-hose water breaker
garden hose
3 two-liter empty soda bottles
1 graduated cylinder
soil
newly-seeded sample
sod
sheet of black plastic
3 bricks for incline
knife

PROCEDURE

1. Cut a V-notch in one end of the greenhouse flat.

2. Line the flat with the black plastic, leaving 3 inches of overlap.

3. Fill each flat with:
   A. field soil
   B. newly-seeded soil
   C. sod

4. Prop up the flats using the bricks and place the two-liter soda bottles underneath the V-notch (cut off the tops to be wide enough to collect the erosion water).

5. Using the water breaker and hose, soak the flat for two minutes and record the erosion runoff.

6. Repeat this experiment for samples B and C.

7. Compare and document the erosion runoff in A, B, and C.

Flat A: _____

Flat B: _____

Flat C: _____

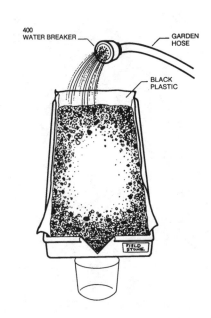

Answer the following questions.

1. Which flat had the most runoff? _____

2. Which flat had the best erosion control? _____

3. What effect does the slope have on soil erosion? _____

_____

_____

4. Explain how this process could be used in the horticulture industry. _____

_____

_____

_____

_____

_____

LAB EXERCISE 2–18

Wetlands Bioengineering

PURPOSE

To stabilize a streambank to help to control soil erosion

Bioengineering is no new practice with technological advances of the future, but rather a practice used in 16th century Europe and still used today in streambank stabilization to help control soil erosion into streams. The streambanks are planted with unrooted branches of native species of black willow, alder, and shrub willows.

This is the bioengineering alternative to the hardscape materials used for streambank stabilization. In this practice, plant materials are used in place of rip rap, concrete liners, gabions, and cribwalls.

MATERIALS

a streambank in need of stabilization
branches of black willow, alder, and shrub willow
biolog made of coconut fiber
$2'' \times 2'' \times 12''$ stakes to fasten the biolog at the stream edge
involvement with Soil Conservation Service and Natural Resources Department of the local area

PROCEDURE

1. Through the professional expertise of the Soil Conservation Service and Natural Resources Department, select a site.

2. Collect samples of natve species of plant material to be used.

3. Fasten the biolog of coconut fiber to the streambank to control erosion and to stick the cuttings of the native species through along the edge of the bank.

4. Stick additional cuttings of the native species along the bank behind the biolog for additional stabilization.

5. Monitor the growth of the native species to observe the erosion control along the streambank.

LAB EXERCISE 2–19

Calculating Soil Excavation

PURPOSE

To calculate soil excavation

MATERIALS

pen or pencil

PROCEDURE

The horticulture industry involves moving topsoil from time to time. To find the amount of topsoil to be excavated, you must first determine the square feet in the area (length × width). Multiply the square footage of the area by the depth in inches and divide by 10 inches. (10 will automatically figure in the 20 percent compaction when topsoil is loose.)

$$\text{Formula} = \frac{\text{length} \times \text{width} \times \text{depth in inches}}{10 \text{ inches}} = \text{cubic feet}$$

$$\frac{\text{cubic feet}}{27 \text{ cubic feet}} = \text{cubic yards}$$

Sample problems:

1. $200' \times 200' \times 3'' = \dfrac{120,0000}{10} = \dfrac{12,000 \text{ cubic feet}}{27} = 444.4 \text{ cubic yards}$

2. $23' \times 25' \times 3'' = \dfrac{1725}{10} = \dfrac{172.5 \text{ cubic feet}}{27} = 63.8 \text{ cubic yards}$

Determine the total cubic yards for the following landscape soil excavation problems.

A. $155' \times 125' \times 3'' =$ _____ cubic yards

B. $3' \times 75' \times 4'' =$ _____ cubic yards

C. $12' \times 12' \times 6'' =$ _____ cubic yards

D. $245' \times 560' \times 3'' =$ _____ cubic yards

E. $452' \times 678' \times 4'' =$ _____ cubic yards

LAB EXERCISE 2–20

What's in a Fertilizer Bag?

PURPOSE
To acquaint students with fertilizer nutrients

MATERIALS
bag of fertilizer
bag of cottonseed meal
bag of bone meal
bag of blood meal
urea
ammonium nitrate
20-20-20 soluble fertilizer
Introductory Horticulture, 5th Edition

PROCEDURE
Fertilizer is material applied to growing media (soil) to supply the elements the plant needs for optimum growth.

Commercial fertilizers have macronutrients of three major nutrients—nitrogen, phosphorus, and potassium. These nutrients are represented by the letters N, P, K, respectively. Nitrogen supplies the element that keeps the plant color dark green during the growth period. Phosphorus encourages flower and root development and increases disease resistance. Potassium ensures starch formation, plant disease resistance, and production of chlorophyll.

The fertilizer analysis is the relationship of each macronutrient, N-P-K, to one another expressed in the form of a percentage. For example, fertilizer analysis of 10-10-10 has 10 percent nitrogen, 10 percent phosphorus, and 10 percent potassium.

1. The instructor will provide a bag of fertilizer with the label.

   a. List the plant food elements _____

   b. Record the percentage of the plant food elements _____

   c. Describe the physical features of the plant food _____

2. Using the bag of soluble fertilizer 20–20–20

   a. Mix a sample of the soluble fertilizer at 100 ppm (parts per million) and 200 ppm. Use the manufactures recommendations of the bag for mixing.

   b. Note the color concentration of each.

3. How many pounds of each macronutrient of N-P-K are included in the following fertilizers?

   Example: a 50 lb. bag of 10%-10%-10% fertilizer
            50 lb. × .10 = 5 lb. each of N-P-K

   Using this formula determine the pounds of active ingredient in each the following fertilizers.

POUNDS OF FERTILIZER	FERTILIZER ANALYSIS	POUNDS OF NITROGEN	POUNDS OF PHOSPHORUS	POUNDS OF POTASSIUM
50 lb.	5-10-5			
20 lb.	23-7-7			
100 lb.	0-30-0			
40 lb.	20-20-20			
75 lb.	30-0-12			
60 lb.	12-6-6			
85 lb.	4-12-6			
35 lb.	16-8-4			
59 lb.	32-6-8			
62 lb.	12-32-8			

Answer the following questions.

1. List two organic fertilizers. _____

2. List two chemical fertilizers. _____

3. What element source is supplied by urea? _____

4. If urea has 45% nitrogen, how many pounds of N are in a ton? _____

5. What does 20-20-20 mean? Give the fertilizer ratio. _____

   _____

6. How many pounds of nitrogen are in 5 tons of leaf mold compost (5.0% N) _____

   and 3 tons of sewage sludge (6.5% N) _____ .

7. What is the compound that creates the blue color in the solution of the 20-20-20 soluble fertilizer?

   _____

8. Explain what nitrogen, phosphorus, and potassium do for plants. _____

   _____

Refer to pages 38 to 42 in your text to answer the following questions.

9. Plant food may be divided into two groups. What are they? _____

   _____

10. What is the difference between a major and minor element? _____

   _____

LAB EXERCISE 2-21

Portable Kitchen Composter

PURPOSE

To demonstrate how we can reduce the amount of home products going to the landfill, and to show the role of earthworms in the decomposition of organic matter produced in the home kitchen

MATERIALS

2 plastic two-liter clear soda bottles
hair blow-dryer
kitchen discards (e.g., discarded egg shells and vegetable portions)
red wigger earthworms (purchased from a local fishing bait store)
dead leaves
grass clippings
½ cup of greenhouse medium or soil
water
knife with adjustable blade
black construction paper

CAUTION
Knives are sharp!
Be careful.

PROCEDURE

STEP 1: Select one of the two-liter soda bottles. Using a sharp adjustable-blade knife, cut the top neck of the bottle off where it starts to taper in. FOR SAFETY WHEN CUTTING THIS BOTTLE, LAY IT DOWN IN A SHALLOW CARDBOARD BOX TO KEEP IT FROM SLIPPING (see figure below). Puncture through the bottle, holding it firmly. Turn the bottle in a clockwise direction to cut off the neck.

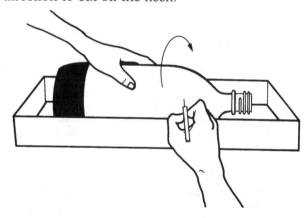

STEP 2: Remove the base of the other two-liter clear soda bottle. Use a blow-dryer to heat the base of the bottle, warming the glue enough to remove the base. This base will be the top cover for your portable kitchen composter:

STEP 3: Add the ½ cup of water to the soil then add the damp soil to the kitchen composter; add a layer of grass clippings and a layer of kitchen discards.

STEP 4: Now put in your earthworms and observe the action. Be sure to keep your composter out of the direct sunlight. Direct light will heat up the composter and kill the earthworms.

STEP 5: Observe the movement of the earthworms after some time. Do you see them working in the composter?

STEP 6: Use a piece of black construction paper to wrap around the bottle. Let this stand for a 24-hour period. What do you observe after this period?

STEP 7: Remember to add additional grass clippings, dead leaves, chopped vegetable portions, and crushed egg shells.

CRUSHED EGG SHELLS

RED WIGGER EARTHWORM

GRASS CLIPPINGS

VEGETABLE PORTIONS

½ CUP DAMP SOIL

Record your observations.

1. Record the action of the earthworms when they were placed in the kitchen composter.

_____

_____

_____

2. Record your daily observations of the action in the composter.

_____

_____

_____

_____

_____

Answer the following questions.

1. How can earthworms help to reduce the tons of garbage that are taken to the landfill?

_____

_____

2. How do earthworms help improve the fertility of soils? _____

_____

3. Are earthworms effective in breaking down organic matter in soil? Explain your answer.

_____

_____

_____

4. What happened when you placed the black construction paper around your kitchen composter?

_____

_____

_____

_____

_____

5. Do you feel it is important to compost? _____ Why? _____

_____

_____

_____

_____

_____

_____

LAB EXERCISE 2–22

Recycling Garden Waste

PURPOSE
To determine how to build a compost pile using garden waste

MATERIALS
4 wooden pallets 3′ by 4′
10-10-10 fertilizer
ground limestone
garden and yard waste (i.e., grass clippings, leaves, vegetable stover, spend flowers, edging material)
food thermometer

PROCEDURE
1. Select an area on school grounds or at home to establish a compost pile. Preferably place it in the service area of the landscape.

2. Set up the wooden pallets on edge and fasten them together with 14 gauge wire at the top and the bottom to construct a bin 3′ high and 4′ wide.

3. Start the compost pile with available organic material (i.e., grass clippings, leaves, straw, hay, chopped tree trimmings). Add about a one-foot depth of material to start.

4. Spread over the area 5 to 6 shovels of topsoil to inoculate the compost pile with the soil microorganisms to start the compost activation process.

5. Spread two cups of 10-10-10 fertilizer over the layer of organic material.

6. Spread one cup of ground limestone over the composting material.

7. Water the compost pile well and mix.

8. Add compost materials as available.

9. Turn the compost pile every week to mix the material and break-down the organic matter.

10. Measure and record the temperature of the compost pile with a soil or composite thermometer.

Week	Temperature	Condition of Material
1		
2		
3		
4		

LAB EXERCISE 2–23

Plant Science
Side 1
Propagation Techniques

Plant Hormones

PURPOSE

To learn how to use rooting hormones to stimulate plant growth

MATERIALS

10 English ivy cuttings
Hormone A
bleach - 10% solution (9 parts water to 1 part bleach)
greenhouse media mix
2 4-inch pots
plant pot labels
propagation bed
pruning shears

> ⚠ **CAUTION**
> Use care when handling
> pruning shears.

PROCEDURE

Select 10 stem cuttings of English ivy (*Hedera helix*) and prepare them for propagation.

1. Collect the cuttings in the field.

2. In the lab, recut the basal stem and remove the lower two sets of leaves.

3. Soak the cuttings in 1% bleach solution (1 part bleach, 9 parts water). Use care in handling bleach.

4. Remove cuttings from the bleach solution after 3 minutes and wash them in clean tap water.

5. Use rooting Hormone A on 5 cuttings (Group A). Place the cuttings in 4-inch pot filled with a greenhouse media mix.

6. Your control group (Group B) will not receive any rooting hormone; stick these cuttings in a 4-inch pot filled with a greenhouse media mix.

7. Place all cuttings in the greenhouse under the mist propagation unit.

8. After 2 weeks, check to evaluate the root growth of the treated (Group A) and the control (Group B) cuttings.

9. After 4 weeks, reevaluate the root growth by measuring the root growth of the controlled cuttings and those with the rooting hormone with a ruler.

10. After 6 weeks, reevaluate again. Record your observations below.

	Group A	Group B
2 weeks		
4 weeks		
6 weeks		

Read Unit 5 in your text and answer the following questions.

1. What is the objective of propagating plants from cuttings? _____
_____

2. Is it necessary to use rooting hormones on all plants? Why or why not?_____
_____

3. List two rooting requirements for geraniums and azaleas. _____
_____

4. What conditions are needed for quick root formation? _____
_____

5. What is IAA and IBA? _____

6. Do all plants contain the same amount of IAA? _____

7. How has the development of rooting hormones benefitted the horticultural industry?_____
_____

8. What are rooting hormones mixed with? _____

9. Rooting hormones come in different strengths which include _____, _____,
and _____. What are they used for? _____

10. Why do most rooting hormones contain a fungicide? _____

SECTION 3

Plant Propagation

◆

NAME: _____   DATE: _____

LAB EXERCISE 3–1

Seed Propagation

PURPOSE

To learn to propagate plants from seeds, an important skill in the horticulture industry

MATERIALS

pen or pencil
seeds and seed catalogs
standard-size seeding flat
germination media (from a horticulture supplier)
1 paper towel
leveling board or skew
mist nozzle (or a special propagation bed mist system with bottom heat)

PROCEDURE

The most common method of plant propagation is by seeds, which is sexual reproduction. Seeds have been the basic structure of plant reproduction since the creation of the Earth. In this lab, you will learn to identify seed parts and functions, prepare media for seeding, sow seeds, provide the proper microclimate for germination, watering, fertilizing, hardening off, and proper handling of seedlings before and during transplanting.

1. Select a germination medium from a horticulture supplier.

2. Select the standard-size flat with good drainage and cover the bottom of the flat with one thickness of a paper towel to prevent the media from falling through it.

3. Fill the flat with the germination medium. Use a skew (see figure below) to level the media and firm it into the flat.

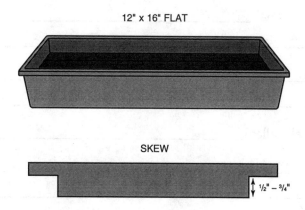

12" x 16" FLAT

SKEW

½" – ¾"

4. Identify the seeds to be planted.

5. Mark rows in the medium for planting the seeds.

6. Open the seed packet and gently sow the seeds into the rows. Be sure to space the seeds. (The "rule of thumb": space seeds the thickness of each seed.)

7. After the seeds are sown into rows, cover the seeds with a thin layer of germination medium. Some seeds will grow easier than others (e.g., tomato and marigold seeds are easy to grow). Caution: use the commercial informational bedding plant guide for proper germination (i.e., to germinate, some seeds need light and others require darkness). Depth of covering will depend on the size of seed. The covering on the seeds should be about the thickness of the seed.

8. It is important to correctly label the seed flats (see below).

Petunia, White Cloud
3-10-82

└ NAME   └ DATE PLANTED   └ VARIETY

9. Keep the medium moist until the seeds germinate by using a mist nozzle or a special propagation bed mist system with bottom heat.

10. Properly store unused materials and clean your work area.

11. Refer back to your seed catalog; it will give you the days to germinate this seed. Compare your results with industry standards.

Refer to pages 59–68 in your text and answer the following questions.

1. What is plant propagation? _____

_____

2. What is the most common method of reproducing flowering plants? _____

3. Define the term "sexual process" _____

_____

4. What is pollen?_____

_____

5. What is the female sex cell called?_____

6. What is self-pollination? _____

_____

7. What is cross-pollination? _____

_____

8. What does "come true to seed" refer to? _____
_____

9. What is a hybrid? _____

10. Define a crossbred. _____

11. What are annuals? _____

12. Label the parts of the seed on the figure below.

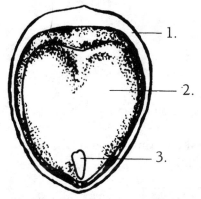

13. List ten annuals easily grown from seed; include their germination times.

_____   _____
_____   _____
_____   _____
_____   _____
_____   _____

14. Draw and label the parts on a germinating seedling.

LAB EXERCISE 3–2

Plant Propagation Word Search

PURPOSE

To develop a vocabulary of sexual propagation of plants by completing the word search
A major portion of the horticultural plants are reproduced by this process.

MATERIALS

pen or pencil
Introductory Horticulture, 5th Edition

PROCEDURE

Find the 14 words in the word search below, using the clues provided.

These words were selected from the material covering sexual plant propagation. There are 14 words hidden in this block of letters. Can you find all 14?

```
m a n q o q a h c c q q v u a
t r m g q l s x w p g l w s a
h e e g p s i c p e c z l t d
b m s p i h f e o d p w g y q
g o s t s t o z b i a v q l f
u d u b a o s i s c l e c e d
p o l l e n d t n e l o x a y
x p f m e z i n q l o x e m f
x g z e c l s j e i r z l e i
z d i o e c i o u s o t u o i
a n t h e r t z i z c q v m s
e e f a g w o y r a v o o v w
n o i t a n i m r e g r g i u
e l c i d a r k c w k q j f n
```

1. The food-storage portion of a seed (Endosperm)
2. The lower portion of the embryo which becomes the primary root (Radicle)
3. The outside part of the seed which protects the seed from injury (Testa)
4. A plant with either staminate or pistillate flowers, but not both (Dioecious)
5. The pollen-producing part of the male sexual organ in the flower (Anther)
6. Tiny spores carrying the male gamate (Pollen)
7. The female sexual structure in the flower (Pistil)
8. The part of the pistil that receives the pollen (Stigma)
9. The elongated part of the pistil between the stigma and ovary (Style)
10. The large lower portion of the pistil where seeds are produced (Ovary)
11. The part of the flower enclosed in the ovary; it becomes the seed (Ovule)
12. The part of the flower composed of petals (Corolla)
13. The base of the flower (Pedicel)
14. The beginning of seed growth (Germination)

LAB EXERCISE 3–3

Plant Propagation Crossword Puzzle

PURPOSE
To complete the crossword puzzle using your knowledge of plant propagation

MATERIALS
pen or pencil

PROCEDURE
Fill in the crossword puzzle by answering the "Down" clues and the "Across" clues provided below.

Down
1. _____ are the off-spring of two different varieties of one plant.
2. At rest.
3. The male sex cell of a plant.
5. Male and female cells from separate parents.
7. The female sex cell of a plant.
8. Lilies can be propagat-ed by _____
10. _____ can be used in plant propagation.
11. Strawberries are repro-duced by _____ .
12. A female plant and a male plant can be _____ed to produce offspring.

Across
4 Some plants can be propagated by the _____ of their roots.
6. Male and female cells are from the same parent.
9. _____ is possible because each single cell of a plant contains the same information.
13. The most common method of reproducing plants.
14. Reproduction of plants.

LAB EXERCISE 3–4

Seed Crossword Puzzle

PURPOSE

To complete the crossword puzzle using your knowledge of seeding, soil amendments, seedlings, and insects

MATERIALS

pen or pencil

PROCEDURE

Fill in the crossword puzzle by answering the "Down" clues and the "Across" clues provided below.

Down

1. You should always _____ your hands before handling seeds or seedlings.
2. The soil _____ can be improved by adding peat-moss.
4. Aphids and mealy bugs.
6. Another word for soil.
7. _____ is used to cover the seeds so that heat will be trapped.
8. Expanded mica.
10. Trees and shrubs are grown here.
11. A small plant shortly after germination.
12. Planting seeds in a location that will allow them to grow without transplanting
13. You _____ the seeds very thinly on the medium.

Across

2. The _____ should be sterilized.
3. Process of moving seedlings
5. C_____ is a method of asexual propagation.
7. Expanded volcanic ash.
9. Seeds starting to grow.
12. The soil may contain a _____ that will kill the seedlings.
13. The covering of a seed.
14. You should always correctly _____ the seed flats after sowing.
15. You should always use _____ when writing a label.

LAB EXERCISE 3–5

Annual Seeds Word Search

PURPOSE

To develop a vocabulary of annual seeds used in the horticultural industry today, by completing the word search Annuals are plants that complete their life in one season.

MATERIALS

pen or pencil

PROCEDURE

Find all of the words listed in the word search below.

Transplant	Seedcoat	Media	Pansy	Germination
Cuttings	Fuchsia	Disease	Rooted	Petunia
Direct	Dustymiller	Growth	Planted	Boston Fern
Structure	Seedlings	Geranium	Soil	Coleus
Germination	Nursery	Felicia	Insect	

```
a f a l g s x t e e r a e n g n h n d q
i o f c x r a r l e i d y e e h p n i b
s g d b m o u i l d k n r e r t o q s e
h h b e c t o l e v a a d l m r h c e e
c j h d c s i m t q n c n l i a q g a o
u k e u j m f e l i c i a y n n w r s k
f e r f y j s i u s e n l c a s o o e h
s t n t e m k m k n g y c q t p w w b b
s v s n a f i v r s s n j u i l j t o b
p u u g p h e e y n e m i w o a h h d h
d e y p r p f c a l l e d t n n i l n a
o u t z e n o p d t k s d q t t r r e c
c e b u o e y r r e l v y l a u j z x v
c k g t n m l b o o t c e r i d c k j g
u v s r u i e y o v h n m v e n r l j x
q o o x o i a o t j r r a z v s g e p y
b s u e l o c l e v d d i l d q r s h k
b f q d d e u a d l o b f m p w g u f n
y l j t l x j t j y i n s e c t r a n o
```

LAB EXERCISE 3–6

Seed Word Match

PURPOSE

To complete the word match using your knowledge of seeds by identifying methods, seed treatment, life cycles, and seed parts

MATERIALS

pen or pencil

PROCEDURE

Match each term with its definition.

1. _____	Endosperm	A.	Result of pollination
2. _____	Fertilization	B.	Expanded mica
3. _____	Annual	C.	First seed leaves
4. _____	Embryo	D.	Plant that lives for more than two years
5. _____	Perennial	E.	Completes its life cycle in one year
6. _____	Seedcoat	F.	Food storage in a seed
7. _____	Germination	G.	Seed begins to grow
8. _____	Scarification	H.	Male and female sex cell uniting
9. _____	Cotyledons	I.	Injury to the seed coat
10. _____	Vermiculite	J.	Outer covering of a seed

LAB EXERCISE 3–7

Seed Germination

PURPOSE

To apply your knowledge of seed germination using various horticultural media, methods, and procedures by answering the following questions completely

MATERIALS

pen or pencil
Introductory Horticulture, 5th Edition

PROCEDURE

Read Unit 6 in your text and answer the following questions.

1. What are the requirements for seed germination?_____

   _____

2. What should soil be composed of for seed germination? _____

3. What type of sand is used for germination? _____

   _____

4. What are four characteristics of sphagnum moss? _____

   _____

   _____

   _____

5. What is perlite? _____

6. What are two reasons for using perlite?_____

   _____

7. What is vermiculite?_____

8. What does a good mix consist of for germination?_____

   _____

9. What is added if the soil is too heavy? _____

10. How is soil pasteurized?_____

   _____

11. What is direct seeding? _____

12. What are seven plants that can be direct seeded?

_____   _____

_____   _____

_____   _____

_____   _____

_____   _____

_____

13. How do you improve soil structure? _____

14. What is done to seeds that have a hard seed coat? _____

_____

15. What are three examples of seeds that have a hard seed coat? _____

_____

_____

16. How do you germinate seeds from apples, peaches, pears, or maples? _____

_____

_____

_____

17. What is a primed seed? _____

_____

18. Why is it advantageous to select seeds that weight the same? _____

_____

19. Why are primed or enhanced seeds best for direct field planting? _____

_____

LAB EXERCISE 3–8

Seed Viability by Seed Germination

PURPOSE

To test seed viability by seed germination
The best possible seed germination means greater profit in the horticulture industry.

MATERIALS

pen or pencil
bean seeds (dicot)
corn seed (monocot)
petri dish
filter paper
distilled water
seed starter
Introductory Horticulture, 5th Edition

PROCEDURE

1. Use 10 seeds each of corn seed (monocot) and bean seed (dicot).

2. Line the petri dish with filter paper and moisten with the distilled water.

3. Place the 10 seeds evenly on the filter paper.

4. Using another piece of filter paper, cover the seeds and moisten the paper.

5. Label the sample with the date, name, and the seeds used. Place in the seed starter for germination, at the optimum temperature.

6. Check the seeds daily and add water to the petri dish to deep the filter paper moist.

7. Repeat the process for the second group of seeds.

Answer the following questions.

1. How many days did it take to germinate? _____

2. Draw and label the parts of the monocot and dicot germinating seeds.

3. What percentage of germination did your seeds have? (Divide the number of germinated by the total number of seeds used.) _____

4. Why is seed germination so important to the agricultural industry? Explain. _____

_____

_____

5. What is the optimum temperature maintained for this experiment? _____

6. The _____ is a new plant that is developed as a result of fertilization.

7. The _____ is the food storage tissue which nourishes the embryonic plant during germination.

8. _____ seeds tend to grow faster and produce larger plants.

9. When you buy seeds you should purchase them from a _____.

LAB EXERCISE 3–9

Plant Science
Side 2
Tree Fruits and Nuts

Selecting and Taking Softwood Cuttings

PURPOSE

To learn how to select and take softwood cuttings

MATERIALS

pen or pencil
sharp pruning shears or knife
container for carrying the cuttings
Introductory Horticulture, 5th Edition

PROCEDURE

1. Your instructor will give a demonstration on the specific equipment needed and how to take a softwood cutting.

2. Select a plant for the propagation using the chart on the following page as a guide for the season to properly propagate plants.

3. Using sharp pruning shears or a knife, and a container for carrying the cuttings, take cuttings 4 to 5 inches long. Make the cuts at a 90-degree angle.

4. Be sure the plants are free from pathogens (disease and/or insects) and broken branches.

Review Unit 7 in your text and answer the following questions completely.

1. The most common method of asexual reproduction is _____

_____ .

2. What parts of a plant can be used for propagation? _____

_____

3. What are the types of softwood cuttings? _____

_____

4. What is a callus? _____

_____

5. What is a tissue culture? Explain the procedure. _____

_____

6. What are the nine most commonly used steps of vegetative propagation? _____

_____

7. The optimum time to root forsythia as a softwood cutting is _____.

8. Early July is the best time to propagate _____.

PLANT NAME	ROOTING HORMONE	MONTH TO TAKE CUTTING IN MID-ATLANTIC STATES	CUTTING WOOD MATURITY	REMARKS
Arborvitae* (*Thuja occidentalis*) (*Thuja orientalis*)	Hormodin #2	Oct.-Dec.	hardwood	Wounding cutting helps.
Azalea (evergreen) (*Rhododendron*)	Rootone or Hormodin #1	early July	semi-hardwood	Use only terminal growth.
Barberry (*Berberis deciduous*)	Hormodin #2	July-Sept.	semi-hardwood	
Barberry (evergreen)	Hormodin #2	Nov.-Dec.	hardwood	
Boxwood* (*Buxus*)	Hormodin #2	July or Nov.-Dec.	semi-hardwood or hardwood	
Cotoneaster	Hormodin #2	July	semi-hardwood	
Dogwood (*Cornus*)	Hormodin #3	June-July	softwood	new growth after flowering
Deutzia	Rootone or Hormodin #1	June-July Nov.-Dec.	softwood hardwood	
English ivy* (*Hedera helix*)	Rootone or Hormodin #1	Nov.-Dec.	hardwood	
Euonymus*	Rootone or Hormodin #1	June-July-Aug. Dec.-March	evergreen-semi-hardwood deciduous-hardwood	
Firethorn* (*Pyracantha*)	Rootone or Hormodin #1	July-Oct.	semi-hardwood	
Forsythia	Rootone or Hormodin #1	June-July Dec.-March	softwood in container hardwood, outside	
Holly (*Ilex*)* (evergreen)	Hormodin #3	July-Aug.	semi-hardwood	terminal growth
Hydrangea	Rootone or Hormodin #1	June-July Dec.-March	softwood hardwood	
Juniper (*Juniperus*)*	Hormodin #3	Nov.-Dec.	hardwood	
Lilac (*Syringa*)	Hormodin #2	directly after flowering	very soft wood	Use fungicide.
Mock orange (*Philadelphus*)	Rootone or Hormodin #1	July Dec.-March	softwood hardwood	
Pachysandra*	Rootone or Hormodin #1	Nov.-Dec.	hardwood	
Privet (*Ligustrum*)	Rootone or Hormodin #1	July Dec.-March	softwood hardwood	
Rhododendron	Hormodin #3	July and Sept.-Oct.	semi-hardwood	Wound cuttings; use only terminal growth.
Spirea (*Spiraea*)	Rootone or Hormodin #1	July Dec.-March	softwood hardwood	species: *vanhouttei*
Viburnum	Rootone or Hormodin #1	May-June July-Aug.	softwood semi-hardwood	fragrant types all others
Weigela	Rootone or Hormodin #1	July-Sept. Dec.-March	softwood hardwood	
Wisteria	Rootone or Hormodin #1	July	softwood	
Yew (*Taxus*)	Hormodin #3	Dec.-March	hardwood	

*May be taken in fall and midwinter.

LAB EXERCISE 3–10

Propagating Softwood Cuttings

PURPOSE

To learn how to propagate softwood cuttings by asexual reproduction. Many of the economically important plants are reproduced in this fashion.

MATERIALS

selected plants
hand pruners or knife
rooting hormone
flats media
plant labels
marking pen
containers (buckets)

PROCEDURE

1. Your instructor will demonstrate how to place the cuttings in the propagation bed using a rooting hormone.

2. Prepare the cuttings for sticking into the propagation bed by stripping off the two bottom leaves, causing the scar tissue on the stem. Note that this scar tissue will callus at the node and form roots.

3. Your instructor will select and explain the correct strength of the rooting hormone. Always have an extra container in which to place the rooting hormone. This will prevent contamination of the entire container of rooting hormone.

4. Place the cuttings into a container of water. Now you are ready to dip the cuttings into the rooting hormone. Take them out, one at a time, and dip the stem (where you stripped the leaves) into the rooting hormone. Remove any excess rooting hormone by tapping with your finger.

5. Stick the cuttings into the greenhouse propagation bed or into flats you have prepared for the cuttings. Keep the cuttings in neat rows and spaced evenly in the bed or flat.

6. Cuttings should be planted to a medium depth of 1 inch. The medium gives the cuttings area for root development and support of the plant.

7. Place the cuttings in a propagation bed under a mist system to keep them turgid while the cuttings produce their callus and roots. If a propagation bed is not available, it will be necessary to mist with a mist nozzle to provide enough moisture for the leaves.

8. Clean your area and return all equipment and materials.

LAB EXERCISE 3–11

Plant Science
Side 2
Tree Fruits and Nuts

Removal of Rooted Cuttings

PURPOSE

To learn how to remove rooted cuttings properly from the propagation bed, a valuable technical skill needed by horticulturists (propagators)

MATERIALS

rooting bed or flat filled with rooted tip cuttings
container to place removed cuttings in

PROCEDURE

1. Your instructor will demonstrate the proper procedure for removing rooted cultures.

2. Remove cuttings without damaging them by inserting a hand underneath the root ball and lifting them out of the bed or flat in which they have been rooted.

3. Place cuttings in a container large enough to prevent damage to the plant and in which the cuttings can be kept moist.

4. Store cuttings at the appropriate temperature.

5. Clean you work area.

LAB EXERCISE 3–12

Transplanting Cuttings

PURPOSE

To learn how to transplant cuttings into containers for optimum production, a valued technical skill needed by horticulturists

MATERIALS

cuttings
sterile containers
growing medium

PROCEDURE

1. Identify the cuttings to be transplanted.

2. Fill the sterile containers with medium.

3. Using the thumb, forefinger, or stick, make holes in the medium into which the cuttings are to be transplanted.

4. Transplant cuttings, making sure there is good contact between the roots and stems, into the planting medium.

NOTE: Cuttings should be transplanted at the depth that they were initially growing. Cuttings must be handled with care and transplanted in sterile containers in such a way that 90 percent survive and grow to maturity.

LAB EXERCISE 3–13

Plant Science
Side 1
Biotechnology

Venus' Flytrap Tissue Culture

Reprinted with the permission of Carolina Biological Supply Company, Burlington, NC 27215

PURPOSE

To multiply the Venus' Flytrap (*Dionaea muscipula*) by means of a tissue culture.

Tissue culture is a fast growing method of plant propagation that can produce vast numbers of new plants for the horticultural industry.

MATERIALS

Venus' Flytrap culture*
10 tubes Venus' Flytrap multiplication medium
10 tubes Venus' Flytrap pretransplant medium
2 teasing needles
forceps
2 racks
sterile water in tubes
sterile petri dishes
70% alcohol
paper towels
100 ml graduated cylinder
10 ml graduated cylinder
wastebasket
scalpels

⚠ CAUTION
Scalples are sharp!

ABOUT THE VENUS' FLYTRAP

Venus' Flytrap, *Dionaea muscipula*, is probably the best known of all the carnivorous plants. It is native to southeast North Carolina and a small corner of northeast South Carolina in acid, nitrogen-deficient soil. The diameter of the rosette of leaves is usually 3 to 10 inches. The trap is a modified leaf that is supported by a petiole attached to the rhizome. In full sun, the petioles are wide and flat and lie close to the ground, while those in more shaded areas are thinner and more erect.

Insects may be attracted to the trap by the sometimes reddish color and by the odor from the nectar glands, but more often the insects that are caught just happen to walk in the wrong place. If an insect or object touches two trigger hairs (or one hair twice) about 20 seconds apart, the trap shuts and the guard hairs interlock to trap the victim. If no insect is caught, the trap soon reopens. If an insect is captured, the trap continues to close tighter around the insect, the digestive glands secrete enzymes to digest the insect, and the insect is broken down into products absorbed by the plant. If the insect is too large, the trap cannot digest it and the trap dies. New petioles and traps are continually being produced.

Small white flowers are borne on a scape. The tiny black seeds form and mature in 6 to 8 weeks.

VENUS' FLYTRAP CULTURE

The culture supplied with this kit is a Stage 2 actively multiplying Venus' flytrap. It is ready to be subcultured upon arrival but will, if left alone, continue to multiply and grown until the medium is exhausted. It is best to subculture the cultures every 3 to 4 weeks. This enables them to stay in an actively dividing state; otherwise, there may be a long lag between culture initiation and multiplication.

Stage 1, which is accomplished by taking a portion of a plant from in vivo to in vitro, has already been accomplished. The portion of the plant used to initiate the Venus' flytrap culture was the seed. Seed were surface disinfected and allowed to germinate an agar. Individual plantlets were then placed on multiplication medium to begin Stage 2. Stage 3 involves the production of roots and enlargement of the plant. Stage 4 is the placing of fully formed plantlets in soil.

MEDIA

Venus' flytrap multiplication and pretransplant media are the same as those used for the cape sundew. The formula in the Media Formulation Booklet is therefore the same as for cape sundew. In addition to those ingredients listed, 30 g of sucrose and 8 g of agar have been added for each liter of medium prepared. The medium was adjusted to a pH of 4.5 before the agar was added. It was then heated until the agar dissolved, dispensed into tubes, capped, and autoclaved for 15 minutes at 15 psi (121° C).

Remove the tubes from their containers and place them in the white racks. Leave the bands around the tubes until you are ready to use them. You may wish to store unused media in a refrigerator to help prevent evaporation. Be sure to allow the tubes to reach room temperature before using.

Plan how you will use the media. Place some plantlets on multiplication medium and some on pretransplant medium, and save some of each for subculturing 4 weeks later.

STERILE WORK AREA

Much of the work must be done under sterile conditions. To succeed, have an adequately prepared sterile work area and adhere strictly to sterile technique where called for. Use a horizontal laminar flow unit if possible. Alternately, use a transfer case. If neither of these is available, a temporary transfer case can be made from a corrogated paper box or an aquarium. Set it on its side so the open top faces the culturer. Clean the box well and keep the inside wiped with 70 percent ethanol or 70 percent isopropanol.

Locate your sterile work area away from drafts. Doors and windows may have to be closed. Fans and air conditioners can stir dust and increase contamination. Clean and mop the region around your sterile work area. Always wash your hands with soap and 70 percent ethanol (or other disinfectant) before going to the sterile work area. Wear short sleeves or a clean, lint-free lab coat. Keep hair covered or tied back, as hair is a prime contaminant source. Be careful not to breathe on your sterile items. Try to work out in front of you; it may feel awkward, but it will reduce contamination. Equip the work area each time before you begin, making sure all needed items are within reach.

It is helpful to the culturer and the instructor to "act out" each step before actually beginning. Things needed each time are forceps, teasing needles, ethanol, sterile distilled water, paper towels, and wastebasket.

Instruments can be sterilized by placing them in a jar of ethanol or isopropanol until they are needed. To remove the alcohol, dip the instrument into the sterile water and proceed. This eliminates the need to flame and allows you to keep the work area wet with alcohol. The sterile water should be changed frequently. Have several containers on hand. When finished with an instrument, wipe plant debris on paper towel and return the instrument to the ethanol. Be sure that "dirty" operations are not performed over "clean" or sterile items.

Included in the kit are pair of 10-inch stainless steel forceps for transporting cultures out of and into tubes and teasing needles for separating the plantlets. Keep all instruments covered with 70 percent ethanol or isopropanol for about 30 minutes prior to use. A 100-ml graduated cylinder filled with alcohol is ideal for the 10-inch forceps, and a 10-ml graduated cylinder filled with alcohol is ideal for teasing needles, short forceps, etc. If there is any doubt about sterility, resterilize the instruments.

CULTURING

When you remove the Venus' flytrap from the test tube, you will want to separate it into individual plantlets. To do this, hold one teasing needle in each hand and begin to separate the plantlets. You may be able to see what appears to be a little white "bulb" or several "bulbs" in a row. Actually these "bulbs" are the enlarged bases of the petiole stacked together. The older they are, the longer the string of petiole bases. Black roots form from the petiole bases to anchor the plant. Small roots may be visible on the Stage 2 plantlets. Cultures are separated into individual plantlets and placed on multiplication medium to increase the number of plantlets, on pretransplant medium for root formation, or directly into the soil (see TRANSPLANTING).

New cultures can be started from just one petiole and trap, but the survival rate is low. Those that do survive turn red and little black bumps form on the petiole and trap. Each of these little bumps will grow into a new plantlet.

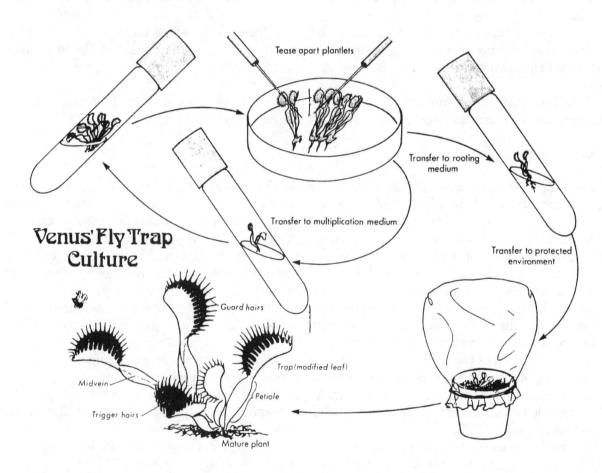

Tease apart plantlets

Transfer to rooting medium

Transfer to multiplication medium

Transfer to protected environment

Venus' Fly Trap Culture

Guard hairs

Trap (modified leaf)

Midvein

Petiole

Trigger hairs

Mature plant

PROCEDURE (Sterile Technique Required)

1. Prepare the sterile work area. You will need the tubed media with bands removed, alcohol, sterile water, forceps, teasing needles, sterile petri dishes or other sterile cutting surfaces, paper towels, and the Venus' flytrap culture.

2. Put forceps and teasing needles in the alcohol.

3. When you begin work do not talk as this can be a major cause of contamination.

4. Place a sterile petri dish in front of you.

5. Remove forceps from the alcohol and dip them in sterile water; remove the flytraps from the tube and place them in the petri dish. Cover the dish and return the forceps to alcohol.

6. Remove teasing needles from the alcohol, dip in sterile water, and separate culture into plantlets. Cover dish while you return scalpels to alcohol.

7. Remove forceps from the alcohol and dip in sterile water. Pick up the tube of medium desired and remove the cap of tube with little finger and ring finger of right hand (reverse if left-handed). Pick up plantlet with forceps and drop in tube; cap tube and return forceps to alcohol. Tube may be tapped or shaken gently to orient plant right side up.

8. Repeat as many times as necessary. Keep records of dates and growth. Watch for contamination. Growth and contamination are easily checked by holding the tube up to the light or by observing under a stereomicroscope.

Cultures should be placed in about 100 foot- candles of light for 16 hours followed by 8 hours of darkness. Temperature should be about 78° F. A bright room is satisfactory. Never place the culture in direct sunlight. If more light is needed, use a fluorescent lamp. Place a thermometer with the tubes to determine the temperature. Cultures can tolerate cooler temperatures, but higher temperatures kill them. If plantlets turns black (not red) or is covered with mold (fuzzy) or bacteria (slick or slimy), autoclave or boil the tube for 15 minutes and discard. Do not try to reuse the medium. The nutrients have been used up and the composition of the medium has been altered.

In 3 to 4 weeks those plantlets placed on multiplication medium can be transferred to either more multiplication medium or to pretransplant medium (see CULTURING). Those plantlets placed on pretransplant medium should be ready to be placed in the soil.

TRANSPLANTING

A mixture of equal parts of sand and peat moss or live sphagnum is recommended. A soilless medium can give good results. Using water (as pure as possible), wet the potting mixture the day before and have it ready. A large plastic pot with drainage holes makes a good culture container.
Rinse the plantlets under lukewarm running water to remove all traces of agar. (The agar medium supports the growth of bacteria and fungi which may kill the tender plantlets.)

Put the plantlets in the pot and place the pot in a plastic bag. A high humidity is necessary until plants can gradually be hardened off. By the second week, start opening the bag for a while, increasing the time each day until the bag can be removed. Do not fertilize until plants are 2 or 3 months old, and then make a 10-fold dilution of what would normally be used. Do not feed the traps. If they "catch" an insect, that is all right, but they can live successfully without it.

* Supplies can be purchased from Carolina Biological Company
2700 York Road
Burlington, NC 27215-3398
Phone: (910) 584-0381
Fax: (910) 584-3399

LAB EXERCISE 3–14

Labeling Plants and Cuttings

PURPOSE
To learn how to label plants and cuttings correctly

MATERIALS
lead pencil or waterproof marker
plant labels

PROCEDURE
1. You are provided with the three plant names below. Correctly print the label with the common name, variety, and date of plantings.

2. Print or type labels legibly, indicating the relevant information for a particular species.

3. Follow a systematic procedure, inserting each label so that the plant it represents is in back of and to its right.

4. Double-check all labels to ensure correct placement.

    LABEL 1 = Poinsettia
               Annette Hegg

    LABEL 2 = Marigold
               Lemon Drop

    LABEL 3 = Tomato
               Better Boy

Complete the labels in the figure below using the current date, e.g., 9/13/96.

LABEL 1

LABEL 2

LABEL 3

SECTION 4

Greenhouse Management and Crops

◆

LAB EXERCISE 4–1

Plant Science
Side 2
Greenhouse Irrigation Systems

Chrysanthemums

PURPOSE

To learn to produce garden mums. This project is an excellent choice for a Supervised Agricultural Experience (SAE) project.

MATERIALS

garden mums from a wholesale supplier
media mix
20-20-20 soluble fertilizer
60-inch azalea pot
plant labels
slow-release fertilizer

PROCEDURE

Garden mums give a great late summer and fall assortment of colors. Garden mums' natural season is fall because the plants are known as short-day plants. Today there are a variety of colors to choose from for the landscape. New flower forms and improved plant quality have made the garden mum very popular with perennial gardeners

NOTE: Fall crops should be started in June.

1. Use 6-inch azalea pots filled with the media mix (e.g., Pro-Mix, Metro-Mix).

2. Place one cutting in the center of the pot.

3. Water and fertilize the mums after transplanting (use 300 ppm of 20-20-20). Continue to feed weekly.

4. Watering is important: do not allow the plants to dry out, as they will wilt, form premature buds, and have reduced branching.

5. Drip tubes or mat systems that supply water by capillary action work well and conserve water; but overhead water may be the only convenient source.

6. To supplement the liquid fertilizer program, use a slow-release fertilizer in the pot.

7. After 2 weeks, pinch the mums to induce branching (breaks) and pinch them again by July 15 in the North and the end of July in the South.

8. Spacing is important: for a 6-inch pot, the spacing should be 14 to 16 inches on center. Proper spacing promotes air movement around the plant.

LAB EXERCISE 4–2

Plant Science
Side 2
Greenhouse Lighting Systems

Zonal Geraniums

PURPOSE

To grow zonal geraniums for spring bedding plants

MATERIAL

rooted cuttings from wholesale growers, 3″ to 4″ inches each
4 2-inch azalea pots
media mix
greenhouse
20-20-20 soluable fertilizer
Epsom salt
Cycocel

PROCEDURE

1. Pot one rooted cutting in a 6-inch azalea pot by March 15.

2. Fertilize with liquid fertilizer with 20-20-20 at 200 ppm.

3. Place in the greenhouse in full sunlight.

4. Give cutting a bi-weekly supplement magnesium (Mg) from Epsom salts (4 oz./2 gallons). Apply 4 oz. of mixed solution of $Mg_2(So4)$ to each geranium.

5. Check the media mix pH and maintain at 6.0 to 6.5.

6. Geraniums are plants that like "warm feet," cool tops, and adequate spacing.

7. Apply Cycocel (1500 ppm) 14 days after transplanting.

8. 7 to 10 days before sale date, reduce liquid fertilizer and water applications to harden off the geraniums.

9. Groom plants by deadheading faded blooms and yellow bottom leaves.

LAB EXERCISE 4–3

Calculating Cubic Air Content of a Greenhouse

PURPOSE

To learn how to calculate the cubic air content of a greenhouse

MATERIALS

pen or pencil

PROCEDURE

Study the diagram, formula, and example shown at the right.

1. Using the formula, calculate the cubic air content of the following greenhouses.

	E	R	W	L
A.	8′	14′	25′	100′
B.	6′	12′	35′	200′
C.	9′6″	16′	55′	125′
D.	10′	18′ 6″	65′	130′
E.	7′	14′	45′	185′

A. _____ cubic content
B. _____ cubic content
C. _____ cubic content
D. _____ cubic content
E. _____ cubic content

2. Measure your own school greenhouse and determine the cubic air volume.

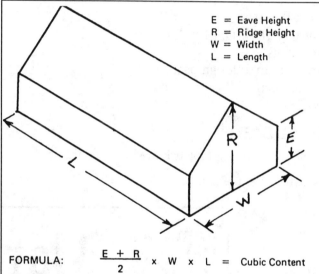

E = Eave Height
R = Ridge Height
W = Width
L = Length

FORMULA: $\dfrac{E + R}{2}$ x W x L = Cubic Content

METHOD: Add E to R and divide by 2. Multiply that answer by W. Multiply this answer by L.

EXAMPLE: E = 6′, R = 12′, W = 20′, L = 100′

$\dfrac{6 + 12}{2}$ x 20 x 100 = Cubic Feet

$\dfrac{18}{2}$ x 20 x 100 = Cubic Feet

9 x 20 x 100 = Cubic Feet
180 x 100 = 18000 Cubic Feet = Content

NOTE; Method can be used only on even span greenhouses.

LAB EXERCISE 4–4

Marketing Signs

PURPOSE

To design a sales sign for the school greenhouse plants to help customers select plants for their needs
This type of signage will improve your sales.

MATERIALS

tagboard paper
markers
rulers
prices of the plants
Introductory Horticulture, 5th Edition
seed catalogs
computer with a design package
plastic protectors to cover signage

PROCEDURE

1. Design a sales sign for use in your greenhouse.

2. See the example below.

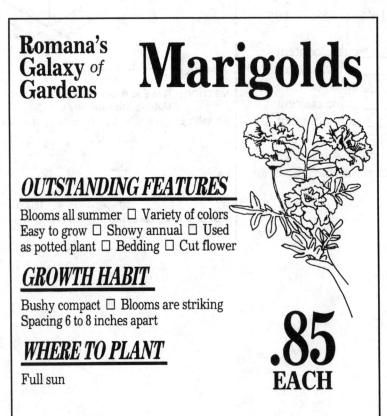

Romana's Galaxy of Gardens

Marigolds

OUTSTANDING FEATURES

Blooms all summer ☐ Variety of colors
Easy to grow ☐ Showy annual ☐ Used
as potted plant ☐ Bedding ☐ Cut flower

GROWTH HABIT

Bushy compact ☐ Blooms are striking
Spacing 6 to 8 inches apart

WHERE TO PLANT

Full sun

.85 EACH

SECTION 5

Integrated Pest Control

◆

LAB EXERCISE 5–1

Plant Science
Side 2
Irrigation and Crop Protection

Anatomy of Insects

PURPOSE
To identify and recognize insect pests of the horticulture industry

MATERIALS
insect collection jar and net
freezer for immobilizing the insects
insect mounting pins
hot plate
250 ml beakers
forceps
½-inch sheets or white styrofoam
scissors
3″ × 5″cards
pen or pencil
Introductory Horticulture, 5th Edition
U.S. Forestry Service Department colored insect chart

PROCEDURE
1. Using the equipment provided for insect collection, scout areas where you suspect insects, e.g., school greenhouse, school grounds, parks, residential sites.

2. Working with your instructor, identify the common insects you find while you scout the assigned areas.

3. Bring the insects to your classroom and observe the different stages of metamorphosis. (The life cycle stages of insects' complete metamorphosis: egg, larvae, pupa, and adult; incomplete metamorphosis: egg, larvae, and adult.)

4. After making those observations, place the insects in the collection jar in the freezer (i.e., standard freezer compartment of a house refrigerator) for twenty-four hours to immobilize them.

5. The next day, remove the insects from the freezer.

6. Using the forceps provided, hold one insect at a time in boiling water for 1 to 2 minutes. This will make the insect more pliable for mounting.

7. Using the mounting pins, hold the insect and put the pin through the right side of its thorax, pushing the pin about three-quarters of the way through. This will keep the insect suspended when it's mounted onto the styrofoam board.

8. Use a 1″ × 2½″ strip cut from an index card (3″ × 5″) for the labeling of the insect. Include the insect name, date collected, location collected, and collector's name.

9. Line the bottom of a soda carrying box for the proper display of your insects.

10. Have a great time scouting for insects!

NAME: _____   DATE: _____

LAB EXERCISE 5–2

Plant Science
Side 2
Irrigation and Crop Protection

Checking Plants for Insects

PURPOSE

To learn how to scout areas to check plants for insects

MATERIALS

insect collection jar and net
freezer for immobilizing the insects
insect modifying pins
hot plate
250 ml beaker
forceps
½-inch sheets whit styrofoam
scissors
information sheet to help recognize the horticultural pests
Introductory Horticulture, 5th edition pp. 169–176

PROCEDURE

1. Read Units 16 and 18 in your text.

2. Your instructor will discuss the parts of the insect, its life cycles, plants, and their effects on control.

3. Observe and identify beneficial and harmful insects (a collection will be provided by your instructor).

4. Make a collection of ten different insects and categorize them as harmful or beneficial. (Use the techniques and procedures in Lab 5–1.)

Answer the following questions.

1. Label the parts of the insect in the figure below.

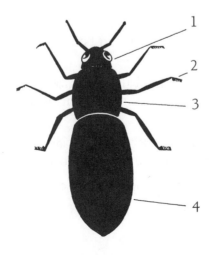

2. List and explain the various methods of how insecticides kill insects.

_____

_____

_____

_____

_____

_____

3. List the different forms that insecticides are sold in.

_____

_____

_____

4. The life cycle of the insect must be understood to control the insect. When are the best times to control insects (include complete and incomplete metamorphosis)?

_____

_____

_____

_____

_____

_____

_____

_____

_____

5. What plants can be used to repel insects? _____

_____

6. What does IPM mean? _____

_____

7. Why is it important that we use biological control of insects? _____
   Explain your answer. _____

_____

_____

_____

_____

LAB EXERCISE 5–3

Plant Science
Side 2
Turfgrass Field Operations

Identifying Weeds

PURPOSE
To identify and recognize weed pests of the horticulture industry

MATERIALS
leaf press
hand trowel
scissors
3″ × 5″ cards
pen or pencil

PROCEDURE

1. Scout the school grounds, parks, and around your home to collect ten different weeds.

2. Using the hand trowel, dig up each weed from the lawn area for your collection.

3. Identify the weeds you collected using an identification chart such as the Ortho Problem Solver, or the Weed Chart from the Lesco Seed Co.

4. Bring the samples back to the lab for mounting. Select a leaf and seed pod for mounting.

5. Using the leaf press, mount samples for pressing. Label each sample with a card: weed name, date, where collected, collector's name.

6. Leave the samples in the leaf press for 4 to 6 weeks. After that time, remove the samples and place them in a three-ring notebook for future reference. Pressed samples must be mounted to paper (in the notebook) with tape or glue.

Answer the following questions.

1. What is a weed? _____

_____

2. What is used to control weeds? _____

_____

3. Explain the difference between a non-selective herbicide and a selective herbicide. _____

_____

_____

_____

4. Develop a one-year weed control schedule for your home.

LAB EXERCISE 5–4

Enticing Wildlife to the Garden

PURPOSE
To construct a bee house

MATERIALS
4″ × 4″ × 8″ cedar, redwood, or pine
electric drill and ⅜-inch wood bit
combination square
screw eye
14-gauge 2-foot wire
saw
sandpaper

PROCEDURE
1. Measure down 3 inches from the top of the timber. Find the center of the timber and draw a line from the 3-inch mark to the center of the timber. Repeat the process for the other side.

2. Using a saw, cut the angle for the top of the bee house and sand any rough edges.

3. Select one side; measure down 4 inches from the top and mark a line across the timber. Move down the timber 1 inch and mark another line. Repeat this process twice so there will be four lines.

4. Using the square, draw three lines at right angles to form a grid at 1 inch apart.

5. Where the lines cross, drill a ⅜-inch wide by 2-inch deep hole. Drill a total of 12 holes. Wear your safety glasses.

6. Screw the screw eye into the top of the bee house.

7. Attach the 14-gauge wire on the bee house.

SECTION 6

Container Grown Plants

◆

LAB EXERCISE 6–1

Planting Container Plants with the Butterfly Technique

PURPOSE

To learn how to plant container plants using the butterfly technique to ensure the proper root growth development in landscape plants

MATERIALS

nursery spade
root-bound container plant
knife

PROCEDURE

1. Remove the plant from the container. The plant may be difficult to remove because of root growth.

2. Use a knife or nursery spade to cut through the bottom half of the root ball.

3. Butterfly the root ball to bring the root system closer to the surface. (Your instructor will demonstrate the butterfly techniques.)

4. Plant shrubs 1 to 2 inches above the existing grade.

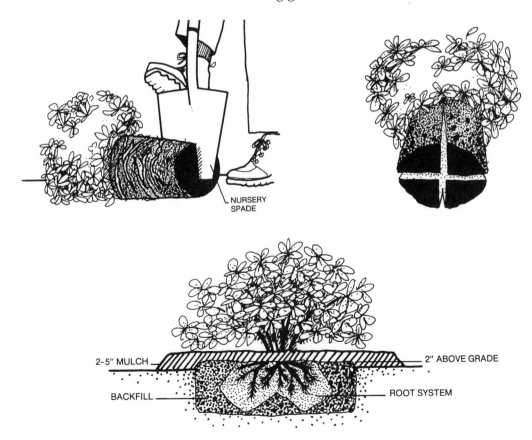

NAME: _____ DATE: _____

LAB EXERCISE 6–2

Creating a Bonsai

PURPOSE

To design and maintain a bonsai planting

MATERIALS

potted plant (e.g., *Serrissa fortida*, *Juniperous procumbens* 'Nana')
bonsai container
9- to 20-gauge copper wire for shaping the plant
screen wire or weed barrier fabric to cover drain holes
potting medium and small gravel
sharp pencil or chopsticks
hand pruners
Introductory Horticulture, 5th Edition
watertight container pan about 4 inches deep
Plant Thrive

PROCEDURE

Review pages 209 to 216 of your text and follow the steps below.

1. Select the potted plant to be trained to bonsai.

2. Select and prepare the bonsai container.

3. Prune to remove unwanted branches.

4. Wrap the copper around the branches to train them in the desired direction by creating movement of the branches. The gauge of copper wire you select will vary depending upon the size of the branches.

5. Using a sharp pencil or chopsticks, remove the soil from around the roots. While removing the soil, look for the center root of the plant and remove.

6. Spread some small gravel over the bottom of the bonsai dish. Pull the wire down through the drainage holes and wire together.

7. Work the medium in and around the plant roots, using the chopsticks to firm the medium.

8. Water your bonsai creation by setting the bonsai into the watertight pan of water with Plant Thrive (B-12 vitamin) to manufacturer's recommendations. Allow bonsai to set in the pan until the bonsai mix is moist to touch. Water the bonsai when soil is dry to the touch.

LAB EXERCISE 6–3

Creating an Interiorscape

PURPOSE

To design an interiorscape that will enhance interiorscape of your school

Employment opportunities are expanding in this area of landscaping.

MATERIALS

selected interior plants of the designer's choice
plant medium
light meter
soil test kit
water test kit
thermometer
graph paper/pencil
Introductory Horticulture, 5th Edition

PROCEDURE

1. Analyze and test the following aspects of your interiorscape location.
   A. light availability
   B. placement of planter boxes (and/or individual containers)
   C. pH level of the soil (medium) and the water
   D. temperature

   Record your results below.

   Light meter readings _____

   Water and soil tests _____

   Temperature _____

2. Using the information you gathered, design an interiorscape, selecting plants best suited to the conditions. When planting the bed, leave the plants in the containers especially during replacement, to remove for pest control.

3. Lay out the placement of the plants on graph paper. This is the design stage.

4. Create your interiorscape: put the plants in their proper places, according to your design on the graph paper.

LAB EXERCISE 6–4

Creating a Totem Pole

PURPOSE

To create a decorative interior planter with vine-like growth.

MATERIALS:

Lab exercise 6–4
interior vine plants (e.g., pothos, philodendron, English ivy)
6-inch Azalea clay pot
plant media
1″ × 1″ × 36″ stake
sphagnum moss
22-gauge wire and/or string
hand pruners
plant shine to clean the leaves
hairpins
Introductory Horticulture, 5th Edition

PROCEDURE

Review pages 221 through 222 of your text and perform the steps below.

1. Select a vine plant.

2. Select a pole that is three to five times taller than the container in which the totem is to be used.

3. Soak the sphagnum moss in a bucket of water.

4. Pack the soaked sphagnum moss around the pole. As you pack the sphagnum, secure it with 22-gauge wire and cover the entire pole.

5. Plant the pothos in the 6-inch pot and firm the medium into place.

6. Set the totem pole in place.

7. Wind the vine around the totem pole.

8. Using hairpins, fasten the vine to the totem pole.

9. Clean the leaves with plant shine.

10. Keep the moss damp; the roots will grow into the moss and the leaves will form a solid mass.

LAB EXERCISE 6–5

Creating a Topiary

PURPOSE
To make a simple topiary, one of the oldest means of creating decorative elements for horticultural display

MATERIALS
rooted cuttings of English ivy (*Hedera helix*)
14-gauge wire
6-inch azalea pots
clay pot
medium
clear fishing line, paper-covered wire, or green yarn
hand pruners

PROCEDURE

1. Make an ivy ring out of the wire as shown below.

2. Fill the azalea pot with soil and insert the ends of the wire in the pot.

3. Plant the ivy cuttings at each end of the wire.

4. As the cuttings grow, wrap them up and around the wire. Tie the cuttings on the wire with florist tape.

5. In time, you will have a lush green ring of ivy—a topiary. Use hand pruners to prune new growth to shape the ivy to the ring.

LAB EXERCISE 6–6

Creating an Herbal Topiary

PURPOSE

To design an herbal topiary using rosemary to create a germometric sculpture for the garden and home

MATERIALS

3-inch pot of rosemary
4-inch clay pot
bamboo stakes
plant ties
hand pruners
scissors
potting medium

PROCEDURE

1. Select a rosemary plant for the herbal topiary design. Note that rosemary has a tendency to produce several branches; you want to select one with a strong main leader to form the base of the topiary.

2. Repot the rosemary into a 4-inch clay pot using the greenhouse medium or soil. Place a crock (a broken piece of clay pot over the bottom) drainage hole to hold in the media. This will allow for good drainage in the pot. Firm the root ball into the pot to assure that the plant will stake and prune well.

3. After selecting the main leader, use the hand pruners to remove the side shoots. (Note the aromatic scent of the rosemary.) At the terminal end (top), leave several branches of the rosemary and prune into a circular shape.

4. Scissors may be used to give a finer cut to the texture of the topiary. Also, scissors may be used to trim any new branches that come out.

5. Save the cuttings from the rosemary to start new plants for other topiary designs.

6. Stake the topiary with a bamboo stake by pushing the stake down beside the main leader into the root ball to the bottom of the 4-inch pot. Tie the main leader to the bamboo stake with a plant tie for support.

7. Water the topiary and watch the rosemary grow.

8. The pruning process should be repeated every 3 to 4 weeks to maintain the design shape.

LAB EXERCISE 6–7

Creating a Window Box

PURPOSE
To construct and plant a window box

MATERIALS
11″ × 6″ × 12′ redwood board
combination square
crosscut handsaw
safety glasses
tape measurer
greenhouse media
16 1¼-inch drywall screws
plants
slow-release fertilizer

PROCEDURE
Window boxes are an excellent way to add color to many areas. Patio, porch, window sill, walkway, foyer, and swimming pool are just a few ideas for use. Attractive window boxes with cascading annuals give a cheery welcome feeling.

1. Build a 1″ × 6″ × 3′ box using the redwood board. Cut three pieces 3 feet long for the bottom and the two sides. Measure the end of the window box and cut the board to cover the end. Wear safety glasses.

2. Assemble the window box using the 1¼-inch drywall screws. Using the combination square check to make sure the box is square.

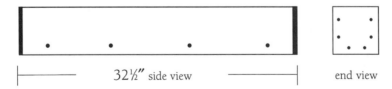

32½″ side view   end view

3. Select a variety of flowering and foliage, cascading and upright plants.

4. Fill the window box with the greenhouse medium.

5. Plant the plants at one half of the in-ground spacing to obtain the instant color desired.

6. Watering and fertilizing are the keys to keeping the window box attractive.

7. Use a complete soluble fertilizer.

8. Regular maintenance is important: pinch by using the thumb and index finger. Pinching will make the plants bushy.

LAB EXERCISE 6–8

Creating a Terrarium

PURPOSE

To design a terrarium

MATERIALS

container (i.e., brandy snifter, rose bowl.)
small potted foliage plants
media mix
sphagnum moss
small stones
activated charcoal
teaspoon and tablespoon
bamboo stakes
florist shears

PROCEDURE

1. Select your terrarium container. Clean it with a glass cleaner and wipe clean with a paper towel.

2. Using a tablespoon, place the small stones in the bottom of the container for drainage.

3. Spread a layer of sphagnum moss over the stones.

4. Using the teaspoon, place a layer of activated charcoal over the sphagnum moss. The activated charcoal absorbs any foul odors in the terrarium.

5. Using the tablespoon, set the plant media mix—be careful not to touch the outside terrarium wall to make it dirty. Add 3 to 4 inches of media mix, firmed down to remove excess water in the mix.

6. Next plant your selected plants in the terrarium. Position the plants for optimum visability.

7. Mist the leaves of the plant material.

8. Cover the terrarium to allow it to establish the natural rain cycle. If the terrarium fogs up, remove the cover to remove excess moisture.

9. Replace the cover to allow the terrarium to establish its natural rain cycle balance.

10. Enjoy your terrarium.

SECTION 7

Using Plants in the Landscape

◆

LAB EXERCISE 7–1

Designing an Annual Flower Bed

PURPOSE

To design and plant an annual flower bed that will give attractive flower color to the selected site

MATERIALS

annuals of the designer's choice
graph paper
scale/ruler
pen or pencil
Introductory Horticulture, 5th Edition
Landscaping by Jack Ingels, Delmar Publishers

PROCEDURE

1. Review Unit 25 in your text.

2. Design an annual flower bed measuring 10′ × 10′.

3. Using graph paper, lay out the boundary lines for the bed.

4. Select the plants from the bedding chart on the following pages. The key factors when designing this flower bed are: color of the flowers and grouping of the colors (see example below), including the mature height of the flowers, spacing of the plants, and sunlight requirements.

5. Design a striking annual flower bed that is aesthetically eye catching.

6. Make a plant key and materials list with plant name, quantity, color, spacing, and height.

7. Prepare the area to be planted using the information provided in Unit 25.

8. Plant the bed and use it as a beautification project.

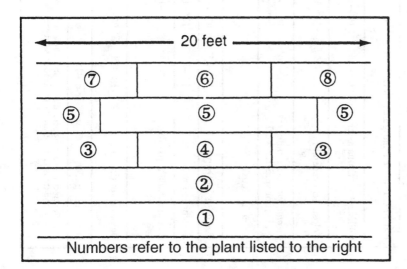

1 Alyssum (snow crystal)
2 Ageratum (blue puffs)
3 Dianthus (rosemarie)
4 Salvia (red hot sally)
5 Celosia
6 Lisanthus
7 Aster
8 Snapdragon

NAME	HEIGHT	SPACING	COLOR	USES	HOW TO START	REMARKS
Ageratum (ageratum)	6"–12"	6"–9"	blue, pink, white	edging	seed and transplant**	Compact, excellent bloom. Needs full sun.
Antirrhinum (snapdragon)	12"–36"	6"–8"	yellow, pink, white, blue, red, salmon	cut flowers, flower bed	seed and transplant	Must be staked for straight flower spikes. Needs full sun.
Aster	8"–36"	10"–12"	blue, red, salmon, pink, white, yellow	flower beds	seed and transplant	Excellent cut flowers.
Begonia (begonia)	6"–12"	6"–10"	pink, red, white	edging, potted plant, window box, hanging basket	cuttings or seed	May be used as a houseplant. Needs direct light.
Browallia (browallia)	12"–18"	8"–10"	blue, white	hanging basket, window box	cuttings or seed	Makes an attractive houseplant. Needs full sun.
Calendula (potted marigold)	12"–24"	10"–12"	yellow, orange	flower bed, hanging basket	seed and transplant	Flower petal used in cooking stews to add color. Needs full sun.
Callistephus (China aster)	12"–24"	10"–12"	blue, purple, white, yellow	flower bed, cut flowers	seed and transplant	Gives excellent summer color. Needs full sun.
Celosia (cockscomb)	12"–30"	10"–12"	red, orange, yellow, pink	flower bed, cut flowers	seed and transplant	Is excellent dried flower. Needs full sun.
Cleome speciosa (spider plant or spiderflower)	24"–36"	18"–36"	pink, white	flower bed	seed directly*	Makes an attractive houseplant.
Coleus (coleus)	12"–24"	10"–12"	red, bronze, yellow, pink foliage	flower bed, hanging basket	seed and transplant	Needs full to partial sun. Beautiful foliage plant.
Coreopsis (coreopsis)	12"–24"	6"–10"	yellow, red	flower bed	seed directly	Good for cut flowers.
Cosmos (cosmos)	48"–72"	12"–18"	red, pink, yellow	flower bed	seed and transplant	Keep near back of beds.
Dahlia variabilis (dahlia)	12"–24"	12"–15"	red, pink, yellow rose	flower bed, cut flowers	seed and transplant	Profuse bloomer. For maximum bloom, sow several weeks before other annuals.
Delphinium (larkspur)	18"–24"	8"–10"	blue, pink, white	flower bed, cut flowers	seed and transplant	Grows best in peat pots. Difficult to transplant.
Dianthus (pink)	6"–18"	6"–8"	white, pink	edging, cut flowers	cuttings or seed	Very fragrant flowers.

* seed directly—sow the seed directly into the soil where the plants are desired.
** seed and transplant—start the seed indoors (in greenhouse, hotbed, or portable seed germination case). As the seedlings develop, they are transplanted from the seed flat to other containers where they grow until ready to be set out.

NAME	HEIGHT	SPACING	COLOR	USES	HOW TO START	REMARKS
Dimorphotheca (cape marigold)	12"–18"	8"–18"	yellow, orange, white	flower bed	seed and transplant	Grows well in dry areas.
Gaillardia (gaillardia)	12"–24"	8"–10"	red, orange	cut flowers, flower bed	cuttings	Loves seashore conditions. Does well in dry areas.
Gazania	6"–10"	8"–10"	orange, yellow, white, tangerine, bronze	flower bed	seed and transplant	Bright blooms.
Gomphrena (globe amaranth)	9"–24"	8"–10"	blue, pink, white	cut flowers, flower bed	seed and transplant	Makes good dried flower. Collect and hang in dry, dark place.
Gypsophila (baby's breath)	12"–18"	10"–12"	pink, white	flower bed, cut flowers	seed directly	This is the annual form; there is also a perennial plant.
Helianthus (sunflower)	18"–60"	12"–36"	yellow, brown	flower bed	seed directly	Seeds are edible.
Helichrysum (strawflower)	24"–36"	8"–10"	yellow, red, white	flower bed	seed and transplant	Makes good dried flower. Cut; hang in dry, dark place.
Hypoestes	8"–24"	10"–12"	burgundy, red, rose, white, pink	hanging baskets, window boxes, flower bed	seed and transplant	Foliage accented by splashed colored leaves.
Iberis (candytuft)	10"–18"	6"–8"	white, pink, red	edging	seed and transplant	Makes attractive ground cover.
Impatiens (impatience)	6"–18"	12"–18"	pink, red, white, multicolor	edging, hanging basket	seed and transplant	Does best in shaded area.
Ipomoea (morning glory)	60"–120"	10"–12"	blue	trellis	seed directly	Vine.
Lantana (lantana)	12"–36"	2"–15"	yellow, blue, red	hanging basket, flower bed, potted plant	cuttings	Makes excellent topiary plant for patio.
Lathyrus (sweet pea)	36"–60"	6"–10"	pink, blue, white	cut flowers	seed directly	Grows best in cool conditions.
Limonium (statice or sea lavender)	18"–30"	10"–12"	blue, pink, white	flower bed, cut flowers	seed and transplant	Makes good dried flower.
Lisanthus	6"–24"	12"–14"	blue, rose, orchid, white, pink, yellow	flower bed	seed and transplant	Large blooms, drought-tolerance.
Lobelia (lobelia)	3"–6"	4"–6"	blue, pink, white	edging, hanging basket, potted plant	seed directly	Very attractive in hanging basket.
Lobularia (sweet alyssum)	4"–6"	4"–6"	white, pink, blue	window box, edging, hanging basket	seed and transplant	Blooms all summer.

NAME	HEIGHT	SPACING	COLOR	USES	HOW TO START	REMARKS
Matthiola (stock)	12"–36"	8"–10"	white, pink, yellow, red	cut flowers, flower bed	seed and transplant	Fragrant flowers.
Mirabilis (four-o'clock)	24"–36"	10"–12"	yellow, red, white	flower bed	seed and transplant	Withstands city conditions.
Myosotis (forget-me-not)	6"–12"	6"–8"	blue, pink	edging, cut flowers	seed directly	Makes attractive cut flowers.
Nicotiana (flowering tobacco)	18"–24"	10"–12"	rose, white	cut flowers, flower bed	seed and transplant	Fragrant flowers after dark.
Papaver (poppy)	18"–36"	10"–12"	red, yellow, white	cut flowers, flower bed	seed directly	Grows in masses.
Pelargonium (geranium)	12"–18"	10"–12"	red, white, pink, lavender	flower bed, window box, hanging basket	cuttings and seeds	Excellent flowering plant.
Petunia (petunia)	6"–18"	10"–12"	pink, red, white, blue	edging, flower bed, hanging basket	seed and transplant	Most widely used annual.
Phlox (phlox)	6"–18"	6"–8"	pink, blue, white, red	cut flowers, flower bed, window box, hanging basket	seed directly or transplant	Very intense colors.
Portulaca (rose moss)	4"–6"	3"–6"	red, orange, yellow	edging, hanging basket	seed directly	Reseeds by itself.
Salpiglossis (giant velvet flower)	18"–36"	10"–12"	gold, scarlet, rose, blue	flower bed, cut flowers	seed and transplant	Very small seed.
Salvia (scarlet sage)	10"–36"	6"–12"	blue, red, white	cut flowers, flower bed	seed and transplant	Very showy color.
Scabiosa (pincushion flower)	30"–36"	12"–14"	multicolor	cut flowers	seed and transplant	Variety of plants from which to select.
Snapdragon	12"–36"	8"–10"	pink, white, red, rose, bronze, orange, salmon	flower bed	seed and transplant	Excellent background plants.
Tagetes (marigold)	6"–48"	6"–12"	orange, yellow	flower bed, edging, potted plant, hanging basket	seed and transplant or seed directly	Wide range of varieties. No insect problems.
Thunbergia (black-eyed Susan vine)	24"–60"	10"–12"	orange	climbing vine, hanging basket	seed and transplant	
Tropaeolum (nasturtium)	12"–18"	10"–12"	orange, yellow	flower bed, window box, hanging basket	seed and transplant or seed directly	Flowers and leaves are edible.
Verbena (verbena)	6"–12"	10"–12"	purple, red, white	flower bed, edging	seed and transplant	Good rock garden plant.
Viola (pansy)	6"–8"	6"–8"	multicolor	edging, flower bed, cut flowers, hanging basket	seed and transplant	Gives excellent color in summer.
Zinnia (zinnia)	8"–36"	6"–12"	multicolor	edging, flower bed, cut flowers	seed and transplant or seed directly	Gives excellent color in summer.

LAB EXERCISE 7–2

Designing a Perennial Flower Bed

PURPOSE:

To design and plant a perennial flower bed that will offer an aesthetically pleasing display of color and foliage year after year

MATERIALS

perennials of the designer's choice
graph paper
scale/ruler
pen or pencil
Introductory Horticulture, 5th Edition

PROCEDURE

1. Review Unit 26 in your text.

2. Design a perennial flower bed measuring 10′ ×10′.

3. Using graph paper, lay out the boundary lines for the bed.

4. Select any of the plants listed in the bedding chart on the following pages. The key factors when designing the flower bed are: color of the flowers and grouping of the colors, including the mature height of the flowers, spacing of the plants, and sunlight requirements.

5. Design a unique perennial flower bed.

6. Make a plant key and materials list with plant name, quantity, color, spacing, and height.

7. Prepare the area to be planted using the information provided in Unit 26.

8. Plant the bed and use it as a beautification project.

NAME	WHEN TO PLANT SEED	EXPOSURE	GERMINATION TIME (DAYS)	SPACING	HEIGHT	BEST USE	COLOR	REMARKS
Achillea mille folium (yarrow)	early spring or late fall	sun	7–14	36"	24"	borders, cut flowers	yellow, white, red, pink	Seed is small. Water with a mist. Easy to grow.
Alyssum saxatile (golddust)	early spring	sun	21–28	24"	9"–12"	rock garden, edging, cut flowers	yellow	Blooms early spring. Good in dry and sandy soils.
Anchusa italica (Alkanet)	spring to September	partial shade	21–28	24"	48"–60"	borders, background, cut flowers	blue	Blooms June or July. Refrigerate seed 72 hrs. before sowing.
Anemone pulsatilla (windflower)	early spring or late fall for tuberous	sun	4	35"–42"	12"	borders, rock garden, potted plant, cut flowers	blue, rose, scarlet	Blooms May and June. Is not hardy north of Washington, D.C.
Anthemis tinctoria (golden daisy)	late spring outdoors	sun	21–28	24"	24"	borders, cut flowers	yellow	Blooms midsummer to frost. Prefers dry or sandy soil.
Arabis alpina (rock cress)	spring to September	light shade	5	12"	8"–12"	edging, rock garden	white	Blooms early spring.
Armeria maritima alpina (sea pink)	spring to September	sun	10	12"	18"–24"	rock garden, edging, borders, cut flowers	pink	Blooms May and June. Plant in dry sandy soil. Shade until plants are well established.
Aster alpinus (hardy aster)	early spring	sun	14–21	36"	12"–60"	rock garden, borders, cut flowers	white	Blooms June.
Astilbe arendsii (false spirea) 'Europa' 'Fanal' 'Deutschland' 'Superba'	early spring	sun	14–21	24"	12"–36"	borders	pink, red, white	Blooms July and August. Gives masses of color.
Begonia evansiana (hardy begonia)	summer in shady, moist spot	shade	12	9"–12"	12"	flower bed	yellow, pink, white	Blooms late in summer. Can be propagated from bulblets in leaf axils.
Bergenia purpurascens (bergomot)	late winter	light shade	10	18"	2'–3'	medicinal	pink, red	Hummingbirds love it.

NAME	WHEN TO PLANT SEED	EXPOSURE	GERMINATION TIME (DAYS)	SPACING	HEIGHT	BEST USE	COLOR	REMARKS
Candytuft (Iberis)	early spring or late fall	sun	20	12"	10"	rock garden, edging, ground cover	white	Blooms in late spring. Prefers dry places. Cut faded flowers to promote branching.
Canterbury bells (Campanula medium)	spring to September (Do not cover.)	partial shade	20	15"	24"–30"	borders, cut flowers	white, pink, blue	Divide mature plants every other year. Best grown as a biennial.
Carnation (Dianthus caryophyllus)	late spring	sun	20	12"	18"–24"	flower bed, borders, edging, rock garden	pink, red, white, yellow	Blooms in late summer. Cut plants back in late fall and hold in cold frame.
Cerastium tomentosum (snow-in-summer)	early spring	sun	14–28	18"	6"	rock garden, ground cover	white	Blooms in May and June. Forms a creeping mat and is a fast grower. Prefers a dry spot.
Chinese lantern (Physalis alkekengi)	late fall	sun	15	36"	24"	borders, specimen plant	orange	Lanterns are borne the second year in the fall.
Columbine (Aquilegia)	spring to September	sun or partial shade	30	12"–18"	30"–36"	borders, cut flowers	wide color range	Blooms in late spring. Best grown as a biennial.
Coreopsis lanceolata	early spring	sun	5	30"	24"	borders	yellow	Blooms from June to fall if old flowers are removed.
Daisy, Shasta (Chrysanthemum maximum)	early spring to September	sun	10	30"	24"–30"	borders, cut flowers	white	Blooms June and July. Best grown as a biennial in well-drained location.
Delphinium elatum (delphinium)	spring to September	sun	20	24"	48"–60"	borders, background, cut flowers	blue, lavender, white, pink	Blooms in June. Best grown as a biennial. Needs dry location.
Dianthus deltoides (pink)	spring to September	sun	5	12"	12"	borders, rock garden, edging, cut flowers	pink	Blooms in May and June. Best grown as a biennial. Needs dry location.
Foxglove (Digitalis purpurea)	spring to September	sun or partial shade	20	12"	48"–60"	borders, cut flowers	pink, white, purple, rose	Blooms in June and July. Grown as a biennial. Shade summer plantings.

NAME	WHEN TO PLANT SEED	EXPOSURE	GERMINATION TIME (DAYS)	SPACING	HEIGHT	BEST USE	COLOR	REMARKS
Gaillardia grandiflora (gaillardia)	early spring or late summer	sun	20	24"	12"–30"	borders, cut flowers	scarlet, yellow	Blooms from July until frost.
Gypsophila paniculata (baby's breath)	early spring to September	sun	10	48"	24"–36"	borders, cut flowers, drying	white, pink	Blooms early summer until early fall. Needs lots of lime.
Hemerocallis (daylily)	late fall	sun or partial shade	15	24"–30"	12"–48"	borders, among shrubbery	pink, red	Plant several varieties for longer blooming season.
Hibiscus moscheutos (Mallow Rose)	spring or summer	sun or partial shade	15 or longer	24"	36"–96"	background, flower bed	white, pink, red, rose	Blooms July to September.
Iris	bulbs or rhizomes in fall	sun or partial shade	next spring	18"–24"	3"–30"	borders, cut flowers	blue, red, yellow, pink, bronze, wine	Blooms spring and summer if different varieties are used.
Liatris pycnostachya (gayfeather)	early spring or late fall	sun	20	18"	24"–60"	borders, cut flowers	rose-purple	Blooms summer to early fall. Easily started from seed.
Lupinus polyphyllus (lupine)	early spring or late fall (Soak before planting.)	sun	20	36"	36"	borders, cut flowers	white, yellow, pink, rose, red, blue, purple	Blooms most of summer. Needs excellent drainage. Does not transplant easily.
Menthia ssp. (mint)	late winter	sun	8	15"	6"–3'	culinary	white	Mint flavors.
Peony (Paeonia)	Plant tubers in late fall 2"–3" deep.	sun	variable	36"	24"–48"	borders, cut flowers, flower bed	pink, red, white, rose	Blooms late spring. Difficult to grow from seed.
Phlox paniculata (summer phlox)	late fall or early winter	sun	25 irregular	24"	36"	borders, cut flowers	red, pink, blue, white	Blooms early summer. Color of flower varies.
Phlox sublata (moss phlox)	grown from stolons	sun		8"	4"–5"	borders	blue, red, white, pink	Blooms in spring. Drought resistant.
Poppy, Iceland (Papaver nudicaule)	early spring	sun	10	24"	15"–18"	borders, cut flowers	white, pink, red	Blooms early summer. Does not transplant easily.
Poppy, Oriental (Papaver orientale)	early spring	sun	10	24"	36"	borders, cut flowers	pink, red, rose, orange, white, salmon	Blooms early summer. Does not transplant easily.

NAME	WHEN TO PLANT SEED	EXPOSURE	GERMINATION TIME (DAYS)	SPACING	HEIGHT	BEST USE	COLOR	REMARKS
Primrose (*Primula polyantha* and *P. veris*)	January, in a flat on surface. Allow to freeze; then bring in to germinate.	partial shade	25 irregular	12"	6"–9"	rock garden	white, yellow, pink, red, blue	Blooms April and May. May be seeded in fall.
Pyrethrum roseum (painted daisy)	spring to September	sun	20	18"	24"	borders, cut flowers	various, including gold, pink, and lavender	Blooms May and June. Prefers well-drained soil.
Rudbeckia (*Echinacea purpurea*) (cone flower)	spring to September	sun	20	30"	30"–36"	borders, flower bed, cut flowers	white, pink, red, rose	Blooms midsummer to fall. Shade summer plantings.
Salvia (*Salvia azurea grandiflora* and *S. farinacea*)	spring	sun	15	18"–24"	36"–48"	borders	red	Blooms August until frost.
Sea lavender (*Limonium latifolia*)	early spring	sun	15	30"	24"–36"	flower bed, cut flowers, drying	pink, yellow, mauve	Blooms in July and August.
Sedum spectabile (sedum)	late winter	sun	10	10"	4"–15"	ground cover	pink, white	Fall foliage.
Stokesia cyanea (Stokes' aster)	early spring to September	sun	20	18"	15"	borders, cut flowers	white, blue	Blooms in September if started early.
Sweet pea (*Lathyrus latifolius*)	early spring	sun	20	24"	60"–72"	background	pink, white, purple, red	Blooms June to September. Easily grown as a vine on fence or trellis.
Sweet William (*Dianthus barbatus*)	spring to September	sun	5	12"	12"–18" (Dwarf form also available.)	borders, edging, cut flowers	red, pink, white	Blooms May and June. Very hardy. Needs well-drained soil.
Veronica spicata (speedwell)	spring to September	sun	15	18"	18"	borders, rock garden, cut flowers	purple	Blooms June and July. Easy to grow.

LAB EXERCISE 7–3

Planting Seasonal Landscape Color

PURPOSE

To enhance seasonal landscape color in a flower bed rotation to demonstrate your landscaping skills In today's landscape business many companies have seasonal enhancement divisions.

MATERIALS

spring: bulbs (e.g., daffodils, tulips, crocus, hyacinths)
summer: annuals (e.g., geraniums, pentas, petunias, begonias)
fall: annuals and perennials (e.g., flowering cabbage, ornamental kale, chrysanthemums)
bonemeal (4–8 lb.) or 2 lb. 10-10-10 fertilizer per 100 sq. ft.
soil test kit
organic matter (3 cu.ft compost leaves per 100 sq. ft.)
rototiller and garden tools for planting

PROCEDURE

1. Select a planting site on the school grounds.

2. Draw, to scale, the landscape planting plan for the selected bed. Plan the materials being planted for each of the seasons.

3. Prepare the flower bed on the school ground. Test the soil and add the amendments recommended by the soil test. Add organic matter (i.e., peat moss, pinebark fines, or composted leaf mold).

4. For spring color, plant bulbs in October or November.

   High maintenance approach:

   A. After flowering, cut off flower heads.

   B. Gather the foliage of the bulbs, fold in half, and tie together with a rubber band around the foliage.

   C. Allow the foliage to turn yellow before cutting the leaves off at ground level.

   D. Apply 4–8 lb. bonemeal per 100 sq. ft. or 2 lb. 10-10-10 per 110 sq. ft. to the bed area.

   Natural approach:

   A. After flowering the foliage allow to die back naturally and then cut off the yellow foliage at the base.

   B. Apply 4–8 lb. bonemeal per 100 sq. ft. or 2 lb. 10-10-10 fertilizer per 100 sq. ft. of bed area.

   C. Apply 3 cu. ft. leaf compost per 100 sq. ft.

5. For summer annuals color, plant as follows:

   A. After the bulbs flower in the spring, prepare the beds for summer annuals. Allowing the bulbs to remain in the ground will mean you will have to take extra precaution when working the bed so as not to damage the bulbs.

   B. Rototill the ground with light penetration of the soil to 4 inches.

  C. Select the summer annuals you have designed in the landscape planting plan.

  D. Irrigate the annuals during the summer months.

  E. Maintain the annuals until September.

6. For fall color, plant as follows:

  A. Remove the summer annuals.

  B. Prepare the soil by adding 2# 10-10-10 fertilizer per 100 sq. ft.

  C. Rototill the soil and add fertilizer to the soil.

  D. Plant your fall crop: flowering cabbage or flowering kale (these plants will stand the cold and frost better than mums) or chrysanthemums.

LAB EXERCISE 7–4

Balling and Burlapping With Drum Lacing

PURPOSE

To ball and burlap with drum lacing, a method of holding the soil to the root ball of a transplanted tree

MATERIALS

tree caliper
tape measure
burlap
knife
6d pinning nails
3-ply twine
10 60d spikes

PROCEDURE

Deciduous trees and shrubs are dug by hand in today's landscape industry. These plants can be dug throughout the growing season in spring and fall; some summer digging can be done, but precautions must be taken. Many of the trees and shrubs sold are balled and burlapped, therefore it is important to understand how to "B&B" with drum lace.

1. Select a deciduous tree to be dug.

2. Assume that the tree has a 2-inch caliper, which is measured 6 inches above the ground level.

3. The general rule of thumb is for every inch in diameter of caliper, you will have a 12-inch root ball.

4. The depth for the root ball is 75 percent of the diameter of the root ball.

5. Once you have determined the root ball size, mark it out with nursery spade, cutting with the back side of the spade toward the tree.

6. Remove all of the grass from the top of the root ball.

7. Dig a trench around the root ball, cutting any roots as you are digging.

8. Once you are at 75 percent of the root ball diameter, you should taper in the base of the root ball with the nursery spade, cutting any fiber or tape roots.

9. Cut the burlap to fit the root ball (burlap comes in various sizes but it is usually 40 inches wide.)

10. After the burlap is fitted to the root ball, use the 6d pinning nursery nails to fasten the burlap to the root ball.

11. Go around the root ball, keeping the burlap tight. (Note that the tree is still in the ground.)

12. On top of the root ball, set the 60d spikes to the outer edge of the root ball by placing the spikes 6 inches apart around the top edge. This is the start of the drum lace.

13. With the 3-ply twine, tie a piece at the base of the root ball. Be sure to tie it tight.

14. With another 3-ply twine, 15 feet long tied to the base twine, bring the twine up to the first 6d

spike and loop it around the spike. (Note: these spikes are only used to hold the rope in place during the drum lacing process.)

15. Continue to lace around the root ball until you have gone completely around the root ball.

16. Lace another piece of twine about 3 feet long through the loops formed at the spikes. Use a weaving pattern through each loop; once completed, tie off the rope with a square knot.

17. Pull out the 6d spikes and store them in a safe place for future use.

18. Now you are ready to tighten the side ropes in the lacing process. Tighten the rope by slipping the rope through each loop; continue until you have gone completely around the root ball. Tie this tightened rope to hold the root ball firmly in place.

19. Cut a piece of burlap 4″ × 12″. This burlap will be used to wrap around the tree trunk at the base to protect it from the rope scarring the bark.

20. Tie a 3-foot piece of rope to the top rope on the root ball. Pull the rope across the top of the root ball and wrap around the trunk. At a right angle go to the rope on top of the ball and loop through it and come back to the trunk and repeat the process until a star is formed on top of the root ball. Tie off the rope.

21. Now you are ready to break the root ball loose from the ground. Push on the root ball to break the tree from the ground.

22. Now tie the bottom of the root ball to prevent any loss of the soil from the root ball.

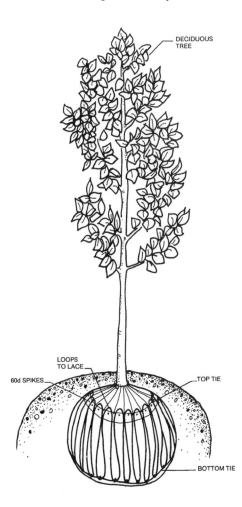

LAB EXERCISE 7–5

Determining Ground Cover

PURPOSE

To determine the needed ground cover

MATERIALS

ground cover materials
prepared bed for planting

PROCEDURE

Ground covers are important because they provide excellent aesthetics, erosion control, and improved beauty of the area. When using ground covers, one must consider plant size availability, quality, and plants needed.

Spacing the ground cover close together will give quick cover, erosion control, weed control, and improved beauty of the area.

How many ground cover plants are needed for the bed area? To figure the number plants needed, multiply the number of square feet by the number of plants required per square foot using the table below.

DISTANCE APART	PLANTS PER SQUARE FOOT
4″	9.1
6″	4
8″	2.25
9″	1.77
10″	1.44
12″	1.0
18″	.44
24″	.25

Answer the following questions using the formula and table above.

1. How many blue rug junipers are needed to cover a bank of 125 sq. ft.? Blue rug junipers should be spaced 18 inches on center (OC)._____

_____

2. For a bed of 5′ × 225′ to be planted in sachysandra, how many plants are needed? Pachysandra should be spaced 8 inches (OC).

_____

3. For a bed of English ivy that has 875 sq. ft., how many plants are needed? English ivy (*Hedera helix*) are planted 6 inches OC. _____

_____

4.

Plant Name	Spacing OC	Bed Area (Sq. Ft.)	Plants Needed
ageratum	10″	135	_____
zinnia	36″	240	_____
marigolds	12″	2200	_____
aster—tall	18″	1680	_____
salvia	12″	525	_____
petunia	10″	1324	_____
ajuga	8″	750	_____
anemone	6″	275	_____
daffodil	6″	300	_____
amarullis	12″	675	_____

5. Determine the retail cost of the plants needed above using the following pricing guidelines:
   Annuals are 6 plants/ $ .89
   Perennials are 24 plants/$19.95
   Bulbs are 6 bulbs/$2.89

ageratum	_____
zinnia	_____
marigolds	_____
aster—tall	_____
salvia	_____
petunia	_____
ajuga	_____
anemone	_____
daffodil	_____
amarullis	_____

LAB EXERCISE 7–6

Designing a Bulb Bed

PURPOSE

To design a flowering bulb bed

MATERIALS

bulbs of the designer's choice
graph paper
scale/ruler
pen or pencil
Introductory Horticulture, 5th Edition

PROCEDURE

1. Review Unit 32 in your text.

2. Design a flowering bulb bed measuring 10′ × 10′.

3. Using graph paper, lay out the boundary lines for the bed.

4. Select the plants listed in bulb catalogues provided by your instructor. The key factors when designing this flowering bulb bed are: color of the flowers and grouping of the colors, the mature height of the flowers, and spacing of the plants.

5. Design a flowering bulb bed.

6. Make a plant key and materials list with name, quantity, color, spacing, and height.

7. Prepare the area to be planted using the information provided in Unit 32.

8. Plant the bed and use it as beautification project.

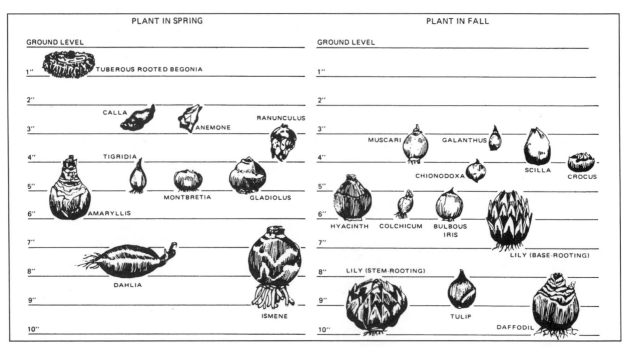

LAB EXERCISE 7–7

Forcing Bulbs for Spring

PURPOSE

To force hyacinth and narcissus bulbs to bloom for Valentine's Day

MATERIALS

precooled bulbs from a bulb wholesaler of hyancinths and narcissi
bulb pans (shallow pans)
planting medium (loam soil or greenhouse medium)
plant labels
rooting rooms to control and regulate temperature and moisture (48° F)
greenhouse for forcing into bloom

PROCEDURE

1. Select the hyancinths and narcissi that will be used for forcing.

2. Buy precooled bulbs from the wholesaler.

3. Plant bulbs into bulb pans in mid-October.

4. Fill the bulb pan to the level that will allow the tops of the bulb to be exposed when planting has been completed.

5. Clean the outer tunic layer (bulb skin) being careful not to damage the basal plate where the roots develop during the cooling treatment. (Note: the skin of hyacinths may cause some irritation to sensitive skin. Wetting your hands may keep this from happening.)

6. When planting the bulbs, place them so they are not touching each other in the pot. For a 6-inch pan you can plant 6 narcissi bulbs and 4 hyacinth bulbs. Work the medium between the bulbs so that only one quarter of the bulb is exposed. This will keep the bulbs from pushing up during the cooling and watering processes.

7. Write up a plant pot label with the date and variety of bulb planted.

8. Place in a rooting room at 48° F for the first 4 weeks. Then lower the temperature to 41° F for the next 10 weeks.

9. Check the soil moisture and rooting room temperature regularly.

10. Move the bulbs out of the rooting room the last week of January for Valentine's Day flowering.

11. Bulbs will need light for 10–12 hours to form bloom buds; grow lights or fluorescent lights can be used.

LAB EXERCISE 7–8

Pruning and Trimming Deciduous Trees

PURPOSE

To learn how to prune and trim deciduous trees properly

MATERIALS

deciduous tree
pruning saw
Introductory Horticulture, 5th Edition

PROCEDURE

1. Review Unit 33 in your text before performing this lab exercise.

2. Your instructor will demonstrate how to use the pruning saw safely and will identify the collar area and branch to be removed.

3. In order to remove large branches from a deciduous tree (i.e., branches that are larger than 1½ inches), you must use the 3-cut method, as in the figure below.

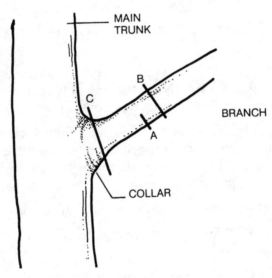

4. Cut A is done about 8 inches away from the main trunk of the tree on the underside of the branch, about one half of the way through the branch being removed. This will prevent the branch from pulling the bark off of the main trunk of the tree when the branch is removed at the collar.

5. Cut B is the second cut; this cut is made all the way through the branch to remove it from the tree about 4 inches away from the main trunk of the tree.

6. Cut C is made at the collar (the rings at the base where the branch attaches to the main trunk). This will allow for better healing of the tree wound. For ecological reasons, pruning paint is no longer used.

LAB EXERCISE 7–9

Pruning Balled and Burlapped Deciduous Trees and Shrubs

PURPOSE

To learn how to prune balled and burlapped deciduous trees and shrubs correctly to reduce the transplanting loss of plant material

MATERIALS

balled and burlapped deciduous tree or shrub
pruning saw
hand pruners

⚠ CAUTION
Use care when handling sharp hand pruners.

PROCEDURE

Pruning should be done on all trees and shrubs to remove dead, broken branches. Except in situations where the plants will be used in a hedge, topiary design, or other specialized pruning practice to enhance the aesthetics of the garden, pruning must maintain the plant's natural shape.

Pruning times are as follows:

• Flowering trees and shrubs that bloom in the spring should be pruned after flowering.

• Flowering trees and shrubs that bloom in the summer should be pruned before flowering (approximately January to April).

• Hedges should be pruned or sheared as needed to maintain the desired shape in the garden.

Pruning procedure:

1. Prune trees just after digging.

2. Thin trees by reducing approximately 30 percent of the crown.

3. Do not cut the main leader.

4. If side branches are cut to balance the tree, make all cuts to the collar on the lateral branch.

5. Final pruning should be done after trees are in place.

BEFORE PRUNING

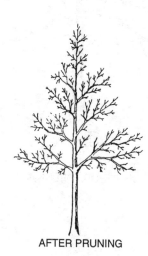

AFTER PRUNING

LAB EXERCISE 7–10

Pruning Bareroot Deciduous Trees

PURPOSE

To learn how to prune bareroot deciduous trees properly to remove any broken, crossing branches before transplanting

MATERIALS

hand pruners

PROCEDURE

1. Prune trees just after digging.

2. Train the top of tree to a strong main leader.

3. Cut the side branches by one third.

4. Remove all sucker growth from the tree.

5. Make all cuts to the collar on the lateral branch or bud.

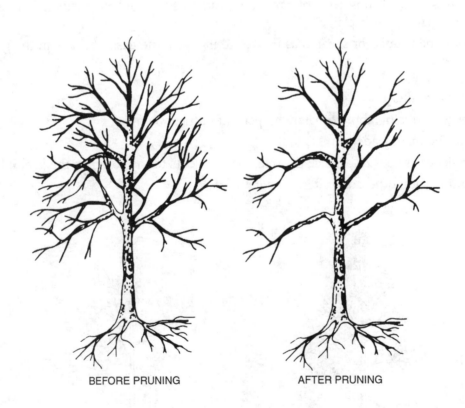

BEFORE PRUNING   AFTER PRUNING

LAB EXERCISE 7–11

Pruning Conifer Trees

PURPOSE
To learn how to conifer trees and shrubs correctly

MATERIALS
conifer tree
shearing knife or head shears
Introductory Horticulture, 5th Edition

Conifer trees are pruned after the new growth has started to harden off. In the case of pines, the best time to prune is June.

The new growth on the pine is called a candle. This candle growth is considered to be one year's growth. In order to maintain a strong pyramidal shape, it is necessary to prune pines by removing 50 percent of the candle each year after the tree is about three years old. It is important never to cut the main leader on conifers. On other conifers such as spruce and fir, pruning is best done in August after the new growth has hardened off.

The spruce and fir differ from pines in their growth habits because they have buds on their branches. These trees branch out easier and fill in more quickly.

PROCEDURE

1. Review Unit 33 in your text.

2. Name the tools required to prune conifers._____

   _____

3. Identify the common name and botanical name of the tree you will be pruning in this lab exercise.

   Common Name _____

   Botanical Name _____

4. What is the best time of the year to prune this tree? _____

5. Prune your conifer to its correct pyramidal shape by removing 50 percent of its candles (remember not to cut the main leader).

LAB EXERCISE 7–12

Using Mulching Beds

PURPOSE:

To learn how much mulch is needed for mulching landscaped beds

MATERIALS

pen or pencil
calculator

PROCEDURE

In the landscape industry today, mulch is used to:

1. control weeds in the beds

2. add organic matter to the soil

3. conserve and hold moisture in the bed area

4. improve the aesthetics of the bed area

There are a variety of mulches used on the market today. Some of the more common mulches used by the landscape industry are pine, bark, shredded hardwood bark, pineneedle straw, coca shells, and cypress bark.

In the spring, apply top-quality mulch comprised of the above materials. The beds should be prepared by removing the weeds and aerating the mulch with a rake or fork. (On shallow rooted plants be cautious not to disturb the root zone.) It is important not to over-mulch because it can have a detrimental effect on the health of plants.

The amount of mulch needed for the landscaped beds must be known when you are bidding on a landscape job. Mulch is applied to the beds from 1½ to 4 inches deep. This will vary with the type of plants that are used in the plan. For example in the case of plants like the rhododendron, it is wise to use mulch in the 1½-inch to 3-inch range.

Mulch is sold in bulk by the cubic yard and bagged by the cubic foot. It is important to know the correct volume of mulch you are using on each landscape job.

Formula:

$$\text{Length in feet} \times \text{width in feet} = \text{square feet (sq. ft.)}$$

$$\frac{\text{Length in feet} \times \text{width in feet} \times \text{depth in inches}}{12 \text{ inches}} = \text{cubic feet (cu. ft.)}$$

$$\frac{\text{Total cubic feet}}{27 \text{ cu.ft./cu.yd.}} = \text{cubic yards}$$

Answer the following questions.

1. Determine the cubic feet of mulch that is needed to cover the following landscape beds using the formula:

   A. 123' × 342' × 4" _____

   B. 32' × 32' × 3" _____

   C. 56' × 78' × 5" _____

   D. 117' × 89' × 4" _____

   E. 147' × 103' × 3" _____

   F. 378' × 639 × 5" _____

   G. 127' × 480' × 3" _____

   H. 1000' × 4' × 4" _____

   I. 198' × 3' × 3" _____

   J. 17.5' × 5.6' × 5" _____

2. Determine the cubic yards of mulch that is needed to cover the following landscape beds using the formula:

   A. 123' × 342' × 4" _____

   B. 32' × 32' × 3" _____

   C. 56' × 78' × 5" _____

   D. 117' × 89' × 4" _____

   E. 147' × 103' × 3" _____

   F. 378' × 639 × 5" _____

   G. 127' × 480 × 3" _____

   H. 1000' × 4' × 4" _____

   I. 198' × 3' × 3" _____

   J. 17.5' × 5.6' × 5" _____

3. Bulk mulch costs 16.75 per cu. yd. when delivered to the job site. Determine the cost of the mulch for A through J above using the following formula:

   Total yards of mulch × Cost per cubic yard = Total cost

   A. _____

   B. _____

   C. _____

   D. _____

   E. _____

   F. _____

   G. _____

   H. _____

   I. _____

   J. _____

4. Lowes Mulch Co. has a tractor trailer to deliver mulch. It has a body 55' long, 8' wide, and 4.5' deep.

   How many cubic yards of mulch can they deliver in one trip? _____

   In A through J which deliveries can be made in one trip? (List by letter.) _____

   _____

   In A through J above, how many trips will each of the other deliveries need? (Round up.)

   A. _____

   B. _____

   C. _____

   D. _____

   E. _____

   F. _____

   G. _____

   H. _____

   I. _____

   J. _____

5. Mulch may be purchased in 3 cu. ft. bags for $2.89 per bag. Determine how many bags of mulch are in A through J (question 1) and how much the total cost would be for each.

A. _____   F. _____

B. _____   G. _____

C. _____   H. _____

D. _____   I. _____

E. _____   J. _____

LAB EXERCISE 7–13

Identifying Power Horticultural Tools

PURPOSE
To identify some of the more common power horticulture tools

MATERIALS
pen or pencil

PROCEDURE

1. Describe the landscape use of the following tools.

Rototiller _____

_____

_____

Power blower _____

_____

_____

Walk-behind mower_____

_____

_____

Chain Saw _____

_____

_____

Line trimmer _____

_____

_____

2. Identify the horticulture tools pictured on the next page.

A. _____

B. _____

C. _____

D. _____

E. _____

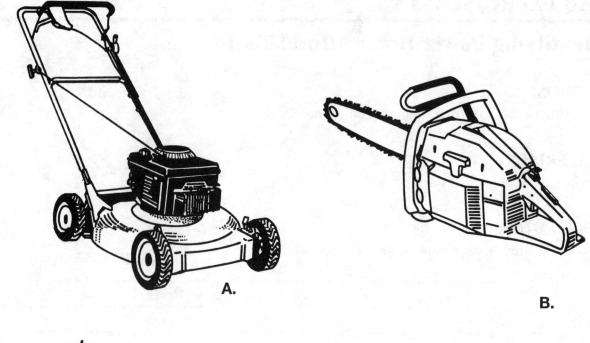

A.

B.

C.

D.

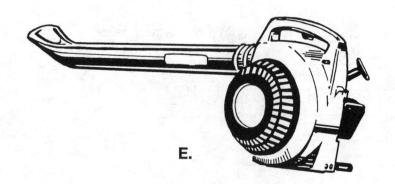

E.

LAB EXERCISE 7–14

Identifying Horticultural Hand Tools

PURPOSE
To identify horticulture hand tools

MATERIALS
pen or pencil

PROCEDURE
The following are important hand tools used in the horticulture industry today. Match the tools pictured on the next page with the correct descriptions below.

_____ A flat spade to edge beds and dig nursery stock

_____ Pole pruners to prune branches high in the framework of the tree

_____ Wheelbarrow to carry mulch and other supplies

_____ Lapping pruner to cut branches 1½ to 2 inches in diameter

_____ 10-tine fork to handle mulch

_____ A special nozzle to water seedling plants

_____ Hand tool to dig pole holes

_____ Hand sprayer for applying chemicals to plants

_____ Tool to shear hedges and topiary forms

_____ Small folding pruning saw

_____ Shovel for digging soil and planting trees

_____ Special wide nozzle water breaker

_____ Hand pruning saw

_____ Flat shovel to remove debris from the landscape

_____ Small trowel to plant seedling plants

_____ A special fork for spading the soil

_____ A small-mouthed water breaker to soften water flow into pots

_____ A special tool to extend your reach while watering plants

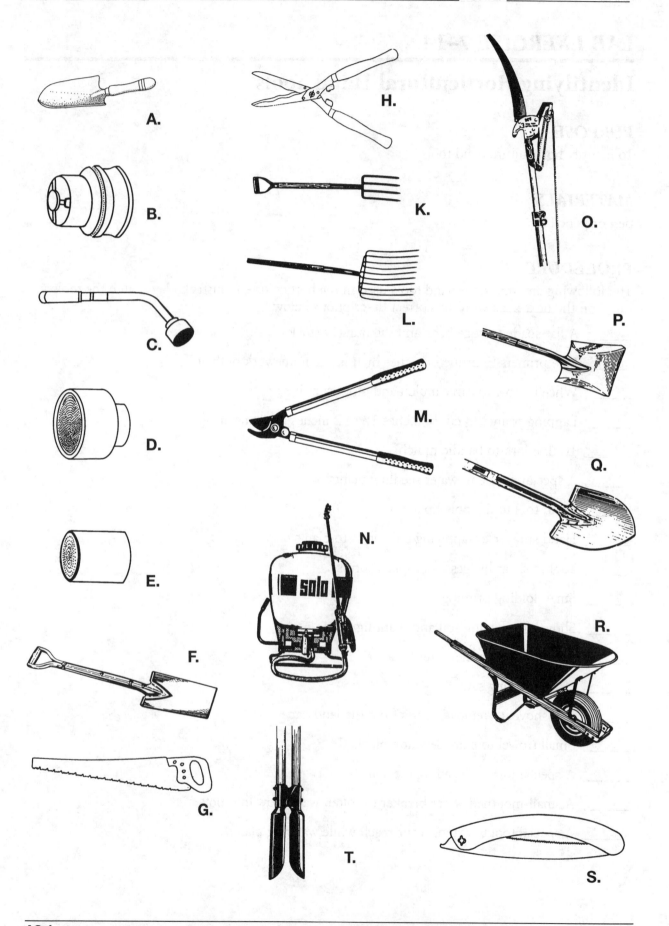

A.

B.

C.

D.

E.

F.

G.

H.

K.

L.

M.

N.

O.

P.

Q.

R.

S.

T.

LAB EXERCISE 7–15

Nursery and Landscaping Word Search (A)

PURPOSE

To develop a vocabulary of scientific and common plant nomenclature

MATERIALS

pen or pencil

PROCEDURE

Find the ten words in the word search.

Benjamin tree	Border forsythia	Chinese holly
English ivy	Ficus	Forsythia
Fraxinus	Hedera	Ilex
Whiteash		

```
e j e w u e n g l i s h i v y z i
c e s h m u p d i u w g w b y m r
i h r h s b q v o a h w o p b d o
f l i t v v g o e m i z t n g i h
d b e n n a w u v e t h r r d t u
k q l x e i r z g f e d r f o j a
o w z q z s m a r e a j a g f t v
h e d e r a e a k g s n v a m q x
r s c s e x x h j h h x f e h h l
y z s v o i d u o n f u i x g e o
v c l a n e g x g l e m h r f w a
e q d u s p x c h r l b p i y x z
u a s n s y b g w k z y c w g g i
d v k f z e f g t k c u c w c n a
j p n w a b h u g g s w f w m h q
b o r d e r f o r s y t h i a w w
```

LAB EXERCISE 7–16

Nursery and Landscaping Word Search (B)

PURPOSE

To develop a vocabulary of scientific and common plant nomenclature

MATERIALS

pen or pencil

PROCEDURE

Find the ten words in the word search.

Adams needle   Canada hemlock   Chinese wisteria
Juddviburnum   Star jasmine   Trachelospermum
Tsuga   Viburum   Wisteria
Yucca

```
c v m t h p k y h a t m b x k y c
s a m m y z o g w d s q h v d i h
l u n q o f o i u a u r z l q x i
w n v a n e e w q m g n j e u a n
u g e h d r w b k s a d z d d f e
g t d w u a l j f n s d o n b n s
t h o g y h w j e t t a a t i e
l m z m s d b e x e k t c a h i w
e j f w n v k x m d p k e c m u i
t w h i t j d r w l i n v c h n s
w m p l s x z m s e o y v u j z t
s t a r j a s m i n e c g y p i e
m u n r u b i v d d u j k l x j r
m u m r e p s o l e h c a r t u i
v h h f m o s m u r u b i v y z a
```

LAB EXERCISE 7–17

Measuring and Drawing Lines with an Architect's Scale

PURPOSE

To learn how to correctly measure and draw lines with an architect's scale

Many of the landscape plans are drawn with an architect's scale.

MATERIALS

architect's scale
pen or pencil

PROCEDURE

Your instructor will introduce and explain the correct use of an architect's scale, and will demonstrate measuring lines using each scale (³⁄₃₂, ³⁄₁₆, ⅛, ¼, ½, 1, ⅜, ¾, 1½, 3).

Using the architect's scale measure the following lines.

A. ├─────────────┤

B. ├──────────────────────────┤

1. How long is each line using scale ¼″ = 1′0″?

   A. _____

   B. _____

2. How long is each line using scale ⅛″ = 1′0″?

   A. _____

   B. _____

3. How long is each line using scale 1″ = 1′0″?

   A. _____

   B. _____

4. How long is each line using scale ½″ = 10″?

   A. _____

   B. _____

5. How long is each line using scale ³⁄₃₂″ = 1′0″?

    A. _____

    B. _____

6. How long is each line using scale ³⁄₁₆″ = 1′0″?

    A. _____

    B. _____

7. How long is each line using scale 1½″ = 1′0″?

    A. _____

    B. _____

8. How long is each line using scale 3″ = 1′0″?

    A. _____

    B. _____

9. How long is each line using scale ¾″ = 1′0″?

    A. _____

    B. _____

10. How long is each line using scale ⅜″ = 1′0″?

    A. _____

    B. _____

Draw a line to scale using the ⅛″ = 1′ and then the ¼″ = 1′ for each of the dimensions listed below. Use the vertical rule as the starting point.

1. 6′

2. 10′

3. 18′

4. 26′

5. 24′

6. 8′

7. 20′

8. 9′

9. 14′

10. 12′

11. 5′

12. 23′

13. 16′

14. 19′

15. 13′

16. 25′ 6″

17. 21′ 9″

18. 1′ 6″

19. 20′ 9″

20. 14′ 6″

LAB EXERCISE 7–18

Measuring and Drawing Lines with an Engineer's Scale

PURPOSE
To learn how to correctly measure lines with an engineer's scale

MATERIALS
engineer's scale
pen or pencil

PROCEDURE
Your instructor will introduce and explain the correct use of an engineer's scale, and will demonstrate measuring lines using each scale (10, 20, 30, 40, 50, 60).

Using the engineer's scale, measure the following lines.

A. |————————————|

B. |————————————————————|

1. How long is each line using the 10 scale?

   A. _____

   B. _____

2. How long is each line using the 20 scale?

   A. _____

   B. _____

3. How long is each line using the 30 scale?

   A. _____

   B. _____

4. How long is each line using the 40 scale?

   A. _____

   B. _____

5. How long is each line using the 50 scale?

   A. _____

   B. _____

6. How long is each line using the 60 scale?

    A. _____

    B. _____

The most common scale used in a landscape drawing is 1″ = 10′; however, in some situations a 1″ = 20′ scale is used. Draw a line to scale using the 1″ = 10′ and then the 1″ = 20′ for each of the dimensions listed below. Use the vertical rule as the starting point.

    1″ = 10′ scale

    1″ = 20′ scale

1. 6′

2. 10′

3. 18′

4. 26′

5. 34′

6. 38′

7. 42′

8. 49′

9. 56′

10. 12′

11. 5′

12. 39′

13. 16′

14. 19′

15. 47′

16. 25′

17. 21′

18. 31′

19. 60′

20. 14′

LAB EXERCISE 7–19

Identifying and Drawing Landscape Symbols

PURPOSE
To learn how to identify and draw landscape symbols use on a landscape plan

MATERIALS
pen or pencil

PROCEDURE
Review the basic symbols recommended for a landscape rendering. The symbols give a graphic view of the type, size, and quantity of plant materials used in the plan. In addition, the plan will have hardscapes, which include walks, fencing, pools, stone, statuaries, and other functional necessities such as utilities (electric, phone, and sewer).

Practice rendering the landscape symbols shown below by drawing three renderings of each landscape symbol.

Shade Trees

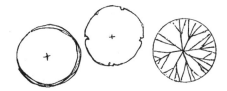

Flowering Trees

Broadleaf Evergreen Shrubs

Azaleas

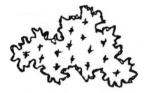

Needle Shrubs

Deciduous Shrubs

Needle Evergreen Trees

Existing Trees

Groundcover

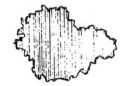

Flagstone

Utility Easement

Soil

Concrete

Wood

LAB EXERCISE 7–20

Using Diameter and Radius

PURPOSE

To learn how to use diameter and radius in landscape plans

MATERIALS

pen or pencil

PROCEDURE

The circle template is a common tool used to render trees and shrubs on the landscape plan. When rendering these plant symbols on the plan, it is important to understand that the symbol represents the mature spread of the plant and the correct planting materials that have to be located on center (OC). Therefore, it is important to understand the parts of a circle: radius, diameter, and circumference.

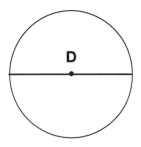 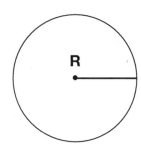

Diameter is the distance across the circle from one side to the other. Diameter ranges from the dripline to dripline of the plant.

Formula: Diameter = 2 × radius

Radius is the distance from the center of the circle to the outer edge. This is important to understand because when plant materials are planted OC it is necessary to measure from the center of the plant.

Formula: Radius = ½ × diameter

SAMPLE PROBLEMS

If the diameter of a white pine tree is 30 feet, what is the radius of this conifer?

r = ½d
r = ½ × 30′
r = 15′

If the radius of a Japanese holly is 12 inches, what is the diameter of this broadleaf evergreen shrub?

d = 2r
d = 2 x 12″
d = 24″

NAME: _____ DATE: _____

Answer the following questions.

Calculate the diameters of the following.

1. Glossy abelia has a radius of 18 inches = _____ diameter

2. Japanese yew has a radius of 4½ feet = _____ diameter

3. Forsythia has a radius of 30 inches = _____ diameter

4. Deodar cedar has a radius of 15 feet = _____ diameter

5. Rosebud has a radius of 15 feet = _____ diameter

Calculate the radius of the following.

6. A patio has a diameter of 16 feet = _____ radius

7. Douglas fir has a diameter of 34 feet = _____ radius

8. Dogwood has a spread of 25 feet = _____ radius

9. Cotoneaster has a spread of 36 inches = _____ radius

10. Pfitzer juniper has a spread of 11 feet = _____ radius

LAB EXERCISE 7–21

Rendering Plants

PURPOSE
To learn how to render plans in a landscape to the scale of ⅛″ = 1′

MATERIALS
circle template or compass
vellum drawing paper
tracing table
pen or pencil

In lab exercise 7–19, you reviewed the symbols used to represent the different landscape and hardscape materials. Using that information, render the correct number of plants into this landscape plan by using a circle template. Render the plants to the correct size as listed (e.g., azaleas 2′ OC—use a ¼-inch circle to represent these plants; 2 feet × ⅛ = 2/8 = ¼).

PROCEDURE
1. Your instructor will explain how to use the circle template and how to render according to the plant's characteristics (e.g., hemlock is rendered as a needle leaf conifer).

2. Place a sheet of vellum drawing paper over the rendering on the next page and trace the outline of the bed area where the plants are to be rendered. Do not sketch the trees freehand; use a circle template or compass. Remember, the scale is ⅛″ = 1′.

3. Render the plants in each of their respective areas using the symbols representing that plant.

4. Keep your drawing neat and clean.

5. When you have completed the rendering exercise create a plant list.

PLANT LIST

Plant Name	Quantity

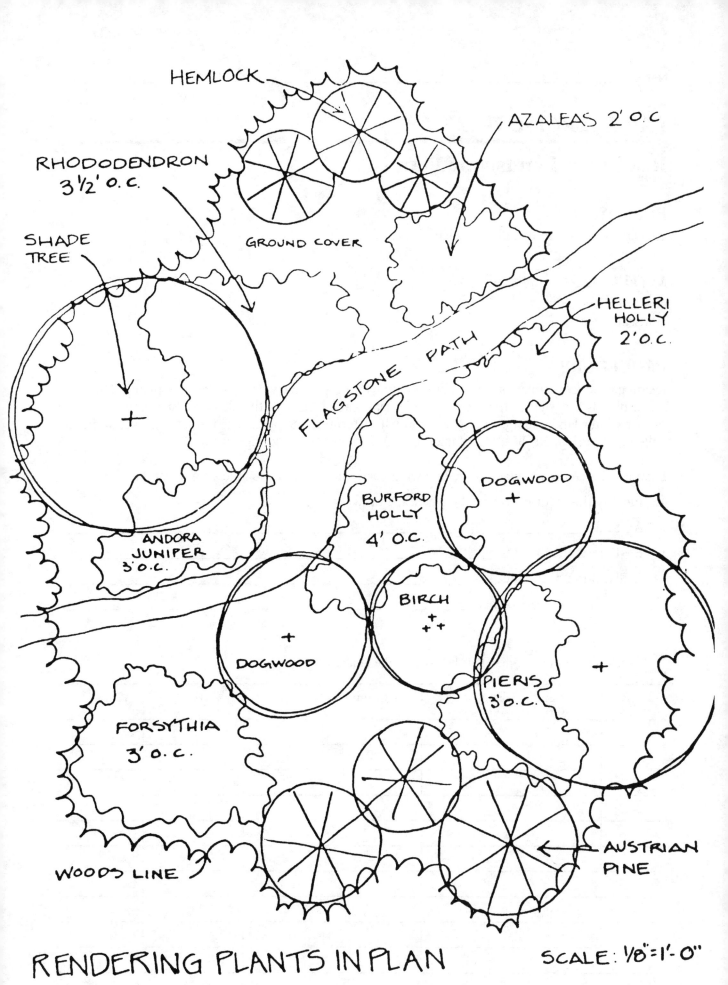

HEMLOCK

AZALEAS 2' O.C

RHODODENDRON
3½' O.C.

GROUND COVER

SHADE
TREE

HELLERI
HOLLY
2' O.C.

FLAGSTONE PATH

DOGWOOD

BURFORD
HOLLY
4' O.C.

ANDORA
JUNIPER
3' O.C.

BIRCH

DOGWOOD

PIERIS
3' O.C.

FORSYTHIA
3' O.C.

AUSTRIAN
PINE

WOODS LINE

RENDERING PLANTS IN PLAN

SCALE: ⅛"=1'-0"

LAB EXERCISE 7–22

Reading a Landscape Plan

PURPOSE
To learn how to read a landscape plan

MATERIALS
pen or pencil
landscape plan

PROCEDURE
Reading a landscape plan is like reading a map; you must know where to place the plants. The land-scape plan is a design of the property to be landscaped. It gives the client a visual concept of the property after landscaping. In this lab, you will determine the plants used, quantity, size, and type: balled and burlapped (B&B), container (cont.), bareroot (Br), rooting cutting (Rc).

Examine the landscape plan on the following page, then answer the following questions:

1. Identify this information on the landscape drawing:

    A. Scale _____

    B. Date of the drawing _____

    C. Location of the site_____

    D. Name of the site owner _____

2. Develop a list of plant materials list used in the landscape plan.

Quantity	Scientific Name (Genus and Species)	Common Name	Size	Type

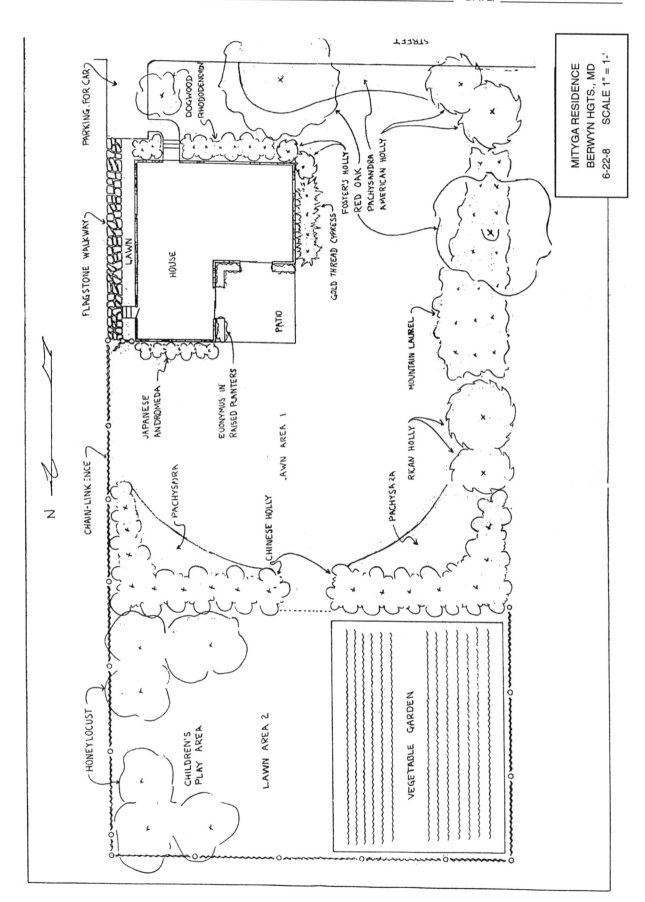

MITYGA RESIDENCE
BERWYN HGTS., MD
6-22-8   SCALE 1" = 1-'

LAB EXERCISE 7–23

Site Planning

PURPOSE
To learn how to classify landscape needs

MATERIALS
graph paper
⅛-inch square
clipboard
100-foot tape
35 mm camera
instant camera
Introductory Horticulture, 5th Edition

PROCEDURE
Your instructor will provide a background of the property to be designed. You will then perform a site analysis of this property. Referring to Unit 34 in your text will be helpful in performing the site analysis, as well.

When you are doing a site analysis, it is necessary to evaluate the plot of land being considered for the development of the landscape plan. Make a gross list of the needs and wants of the client for whom you will develop a landscape plan. This list will determine the capability and desires of the homeowner. It is also wise to take pictures of the property with a 35 mm and Polaroid cameras, which you will use later during the actual landscape drawing.

Ask the following questions when making your list.

1. How many family members are there? _____ What are their ages? _____

   _____

2. How much does the family use the outdoor areas around their home? _____

   _____

3. Does the family entertain frequently? _____ If so, do they primarily entertain large or small

   groups? _____

4. How much privacy from the neighbors and passing vehicles do they want? _____

   _____

5. How much maintenance are they willing to do in the upkeep of the landscape? _____

   _____

6. Are there certain plants they would like to use or that they dislike? _____

7. What services are needed in the landscape (e.g.., storage area, vegetable garden, or compost pile)?

_____

8. Will the garden area be used after dark? _____

9. What kind of budget will you be working with? _____

10. What size plants are needed (i.e., mature specimen plants or smaller landscape sizes)? _____

_____

11. Classify areas in terms of public, service, and private. _____

12. Other notes to take:

- Dimensions of the property _____
- Proposed size of the house (or the current dimensions)_____
- Property building restriction line_____
- Topography of the site _____
- A soil sample to analyze for pH and N,P,K _____
- Location, types, and condition of existing plants (give the exact location) _____

  _____

  _____

  _____

- Location of the utility lines, meters, and easements _____

  _____

- Style of architecture in the area _____
- Good and bad views from the site_____

  _____

- Existing natural resources such as streams, rock outcroppings, specimen plants, and wildlife areas:

  _____

  _____

  _____

LAB EXERCISE 7–24

Water Garden Project Practicum

PURPOSE

To determine the plant material and hardscape materials used in a landscape plan.

MATERIALS

engineering scale
calculator
landscape plans

PROCEDURE

Refer to the landscape plan on the next page. Using the engineering scale and calculator, circle the correct answer for each of the following questions.

1. What is the scale of this plan?
   A. $1'' = 20'$ B. $1'' = 60'$ C. $1'' = 10'$ D. $1'' = 8'$

2. If the line measurement of a line in this drawing were 6½", how long would the line be?
   A. 70'   B. 65'   C. 97'   D. 55'

3. If a line in this drawing were 79' long, how many inches would that measure?
   A. 7"   B. 9"   C. 6½"   D. 7⅞"

4. How many Cornus kousa are used in this plan?
   A. 4   B. 3   C. 6   D. 9

5. What is the mature spread of the Liriodendron tulipifera in this plant?
   A. 15'   B. 30'   C. 12'   D. 50'

6. How many square feet are in the patio area?
   A. 475   B. 500   C. 450   D. 378

7. If you were to use concrete brick pavers on this patio, how many pavers would you have to purchase to complete this project? Assume 45 bricks cover 10 sq. ft. and allow 10 percent breakage.
   A. 2333   B. 2228   C. 1871   D. 2222

8. How many Tsuga canadensis are used in this plan?
   A. 3   B. 5   C. 2   D. 1

9. The garden pond function in this design is
   A. An air conditioner for the courtyard area
   B. A focal point
   C. An irrigation system for the water-loving plants in that area.
   D. A very expensive add-on that contributes little to the landscape.

10. This design was developed with _____ balance.
   A. asymetric   B. informal   C. formal   D. no particular

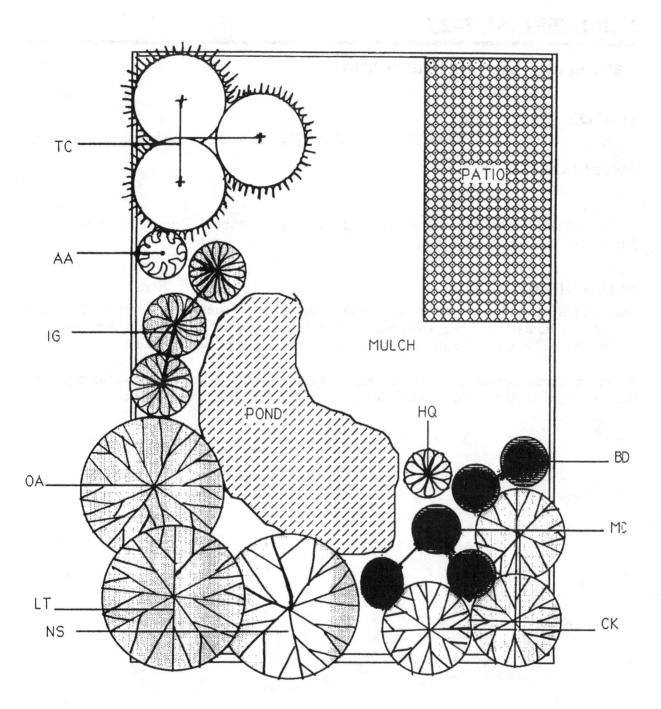

SCALE 1″ = 10′

AA - Agapanthus africanus
IG - Ilex glabra
LT - Liriodendron tulipifera
CK - Cornus kousa
MC - Myrica cerifera

TC - Tsuga canadensis
OA - Oxydendrum arboreum
NS - Nyssa sylvatica
BD - Buddleia davidii
HQ - Hydrangea quercifolia

LAB EXERCISE 7–25

Landscape Sketch Practicum

PURPOSE

To sharpen your skills of landscape sketch by ordering plant materials for the landscape plan

MATERIALS

pen or pencil
calculator
Landscape Plants: Their Identification, Culture, and Use by Ferrell Birdwell, Delmar Publishers
Introductory Horticulture, 5th Edition

PROCEDURE

Research has shown that workers are more efficient and suffer less stress when they are able to view the outdoors from the work site. Today many commercial sites are enhanced with garden areas close to the office building with a landscaped courtyard to provide that tie back to nature.

Review the landscape sketch on the next page. Familiarize yourself with the symbols that represent the plant material. Answer the questions on page 158 regarding this landscape plan.

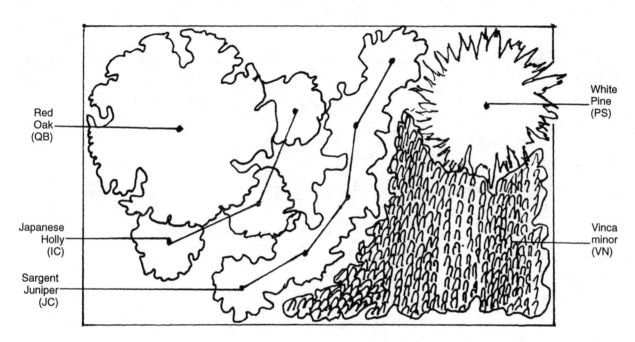

Red
Oak
(QB)

Japanese
Holly
(IC)

Sargent
Juniper
(JC)

White
Pine
(PS)

Vinca
minor
(VN)

Landscape sketch
scale ⅛″ = 1′

N

1. You are responsible for ordering the plant materials for this landscape plan. Make a list of the plant needed, include the key, botanical name, common name, and quantity of plants needed.

KEY	BOTANICAL NAME	COMMON NAME	QUANTITY	REMARKS

2. Test your knowledge of using a landscape sketch and circle the correct answer for each of the questions below.

   A. Based on the scale given, a distance of 28′ should be _____ on the plan.
      a. 1½″   b. 2¾″   c. 3.0″   d. 2.0″

   B. What is the total area of the courtyard?
      a. 1000   b. 1150   c. 950   d. 2590

   C. What is the footage spread of the Red Oak tree?
      a. 32   b. 25   c. 16   d. 40

   D. From the designer's drawing, you would instruct your crew to space JC plants:

      a. 2′ OC   b. 6′ OC   c. 10′ OC   d. 3′ OC

   E. How many IC's are used in this plan?
      a. 10   b. 3   c. 5   d. 7

   F. If this entire courtyard were to be covered with 3″ of mulch how many cubic yards would you buy?
      a. 12   b. 9.5   c. 14   d. 8.0

   G. You can buy bark mulch in bulk at $22.50 per cubic yard. What will it cost to mulch this courtyard?
      a. $213.75   b. $252.50   c. $300.00   d. $455.50

   H. On this plan you measured a line that was 6¾ inches; how many total feet does this represent?
      a. 54′   b. 58′   c. 62′   d. 48′

   I. How many PS is used in this plan?
      a. 5   b. 1   c. 2   d. 7

   J. How many VC's are needed to cover the area if they are planted 8″ on center?
      a. 425   b. 470   c. 690   d. 1050

LAB EXERCISE 7–26

Retainer Walls in the Landscape

PURPOSE
To learn about the various types of retainer walls used in landscape projects

MATERIALS
pen or pencil

PROCEDURE
Read the information provided which explains the various types of retainer walls used. Answer the questions that follow.

INTRODUCTION
For the purposes of our discussion we shall divide the topic of retaining walls shall be divided into three general categories: (1) rip rap walls, (2) cantilever walls, and (3) gravity walls. When the steepness of a slope becomes so great that we cannot use grass or plant materials to stabilize it, we then turn to retaining walls. The first type of retaining wall to consider is rip rap. Actually, rip rap is more of a soil surface-stabilizing situation than a retaining wall. However, we discuss rip rap here because it is a common situation we all encounter and it is possible for the average landscape foreperson to learn the construction techniques necessary to do rip rap work.

RIP RAP WALLS
Below left is a detail of a typical stone rip rap wall. The stones are laid dry (no mortar) and are stacked so that the mass of the stone is transferred to the soil as well as the stone below it. Below right is a typical timber rip rap wall. Note that the timbers are stacked so they overlap about half of the timber below. By doing this, your angle of repose will be approximately 45°. This is important in this type of wall.

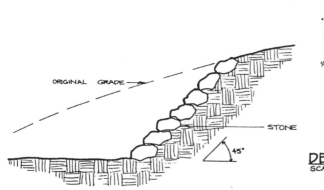

DETAIL 1: STONE RIP RAP WALL
NO SCALE

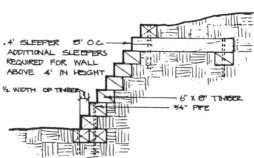

DETAIL 2: TIMBER RIP RAP WALL
SCALE: ½" = 1'-0"

CANTILEVER RETAINING WALLS

Cantilever walls are almost always made of reinforced concrete. The concrete is formed into a "T" or "L" shape. By using this shape, the wall can use the weight of the backfill to help achieve stability. This type of wall is used for retaining from a few feet up to 20 feet. The figure to the right shows a typical example of a cantilever wall. The wall can be faced with brick, stone, or other decorative material. The facing affords no structural stability to the wall. A cantilever wall should be built only by a competent concrete construction contractor. It is not the type of wall the average landscape contractor should attempt to build.

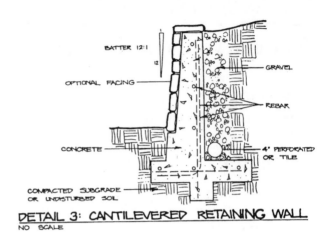

DETAIL 3: CANTILEVERED RETAINING WALL
NO SCALE

GRAVITY RETAINING WALLS

The gravity wall is by far the most common situation the landscape contractor encounters in building retaining walls. The stability of the gravity retaining wall comes from the weight of the wall's mass; simply stated, the wall weighs enough that it won't be easily pushed over from soil and water pressure. The gravity retaining wall should only be used for walls up to 10 feet. Gravity walls can be built in many different ways and with many different types of materials.

CRIBBING

Cribbing can be made of interlocking members of reinforced concrete steel or timbers to form bins filled with selected backfill material to achieve the mass needed for stability. Basically, what you achieve is a "Lincoln Log" effect with concrete or timber.

MONOLITHIC CONCRETE GRAVITY WALL

The figure below shows a typical single poured concrete facing. This wall is somewhat easier for landscape contractors to work with because it requires less critical forming, especially if you were to face the wall with stone or brick. However, it requires lots of concrete, good drainage, and must be separated vertically with control joints.

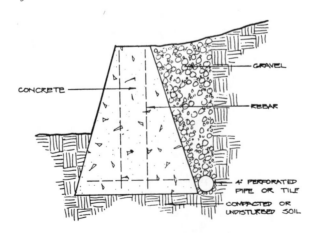

DETAIL 4: CONCRETE GRAVITY WALL

DRYSTONE RETAINING WALLS

Drystone walls use no mortar to bond the stones together. Large, flat stones are used so that they will overlap each other. This overlapping, coupled with the weight of the stone, provides the mass necessary to retain the slope. Drainage is excellent because water can bleed through the wall freely. The wall can also settle unevenly without visual or structural damage. Several things are very important to follow correctly: you must make the base of the wall equal to one half the total height, and you must batter the face of the wall 1 inch for every 12 inches of height. The figure at right shows a detail of how to build a drystone retaining wall.

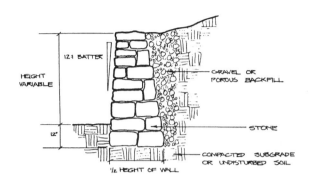

DETAIL 5: DRYSTONE RETAINING WALL
SCALE: 1/2" = 1'-0"

MASONRY STONE GRAVITY RETAINING WALL

With this type of stone wall you still basically rely on the mass of the wall to retain the slope. Therefore, it can be built just like a drystone wall except with mortar in the joints between the stones (see figure to the right). The stone in the rear portion and footing can be of lesser quality than the face stone. Other rubble material could be used as long as it interlocked and bonded properly. One of the most important things to allow for in a solid masonry wall is drainage. The wall can have weep holes with gravel behind them which allow water to bleed through and relieve the pressure. If you have a situation where you don't want the minerals from the stone to wash onto the area below the wall, you must provide a gravel backfill behind the wall and install plastic pipe to carry the water away from behind the wall. You must also provide vertical control joints in a solid masonry wall. This means the vertical bond must be broken so that the wall floats in segments. Many times the footing for this type of wall can be poured concrete which goes below the frost line and is of sufficient width to provide a stable base for the stone masonry.

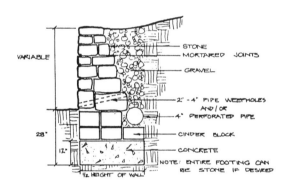

DETAIL 6: MASONRY RETAINING WALL
SCALE: 1/2" = 1'-0"

TIMBER RETAINING WALLS

Timber walls are very similar to cribbing walls. The basic stabilization of this retaining wall is achieved by the use of "deadmen." The deadmen act as footings. They are buried into the hillside and backfill is placed behind the wall. The weight of the backfill actually serves to hold the deadmen down. The deadmen are then attached to the wall and act to hold the wall vertical. This principal is illustrated on the next page. Because wood is an easily amended material, timber can be used in many different ways to build walls. The variations are limited only by your imagination. Caution must be exercised and these basic design principles must be followed:

1. You must stabilize the wall with some type of a deadman.

2. You must provide adequate drainage behind and through the wall.

3. As you go up in height, you must batter the wall more and more. You can achieve this batter by leaning the timbers, thus keeping the timbers level and at the same time achieving the proper batter. The higher the wall, the greater the batter needed, as shown below.

4. Lower walls and single pieces (such as edgings and headers) should be spiked together and pinned to the subgrade and pinned to each other with steel rods, as shown below.

DETAIL 7: TIMBER RETAINING WALL
SCALE: ½" = 1'-0"

DETAIL 8: TIMBER RETAINING WALL

1. What are the three general categories of retainer walls?

_____

_____

_____

2. How are rip rap stone retainers laid? _____

3. What is the normal degree of slope for rip rap? _____

4. What is the recommended height of a cantilever retainer wall? _____

5. What kind of facings can be used on the cantilever retainer wall? _____

6. Who should build a cantilever wall? _____

7. What is the most common wall installed by landscapers? _____

8. What pressures are landscape contractors concerned about when building a gravity retainer wall?

_____

_____

_____

9. What is cribbing? _____

_____

10. What is a monolithic concrete gravity wall? _____

_____

11. How do drystone walls stay in place with no mortar? _____
_____

12. Why is drainage so good on the drystone wall? _____
_____

13. What is the proper batter on a drystone wall? _____
_____

14. Refer to detail of the drystone wall. The batter is 12:1. What does this mean? _____
_____

15. What is the most important thing to consider when you are building a masonry wall? _____
_____

16. What are weep holes? _____
_____

17. If you don't use weep holes, how can you move the water from the back of the wall? _____
_____

18. What is the basic stabilizer of the timber retainer wall? _____
_____

19. List four basic design principles of building a timber retainer wall.

_____

_____

_____

_____

LAB EXERCISE 7–27

Drawing a Construction Detail of a Concrete Gravity Wall

PURPOSE
To draw a construction detail of a concrete gravity wall

MATERIALS
Lab exercise 7-26
pen or pencil

PROCEDURE

1. Using the previous lab exercise, draw the construction detail of the concrete gravity wall

2. This concrete gravity wall is 4 feet high, 2½ feet at the base on a soil foundation, 9 inches wide at the top, has 6 inches of gravel base behind the wall, a 4-inch drainage pipe to remove excess water, and is reinforced with rebar in the concrete.

3. Use the scale $1'' = 1'$.

4. Use the correct landscape symbols and have the measurements accurate, neat, and labeled correctly.

LAB EXERCISE 7–28

Drawing a Construction Detail of a Timber Retaining Wall

PURPOSE
To draw a timber retaining wall detail

MATERIALS
Lab exercise 7-26
drawing equipment

PROCEDURE
1. Using lab exercise 7-26, as a guide, draw a timber retaining wall 4 feet high.

2. Draw the detail and label as shown in the sample.

3. Use the scale ½″ = 1′.

4. Use a 12:1 batter (this means that for each foot of rise, the wall tapers in 1 inch).

SECTION 8

Lawn and Turfgrass Establishment and Maintenance

◆

Plant Science
Side 2
Turfgrass Field Operations

LAB EXERCISE 8–1

Establishing a Lawn Area by Sodding

PURPOSE

To establish a lawn area by sodding

MATERIALS

area to be sodded
certified sod from a local sod farm
soil test kit
sprinkler and rain gauge
rototiller
lawn tools: steel rake, flat shovel, wheelbarrow, round point shovel, nursery spade, lawn roller

PROCEDURE

1. Select an area to be sodded.

2. Prepare the sod bed by rototilling and leveling it with a steel rake.

3. Perform a soil test for pH and N, P, K.

4. Lay the sod in place, in an interlocking pattern (i.e., brick pattern) to decrease the runoff of water.

5. Roll the sodded area with the lawn roller to firm the roots to the soil.

6. Set up a lawn sprinkler and rain gauge. Apply at least one inch of water.

7. Water the sodded area regularly or as needed.

8. The sod should knit to the ground in 8 to 10 days.

LAB EXERCISE 8–2

Determining Total Square Yards of Sod

PURPOSE
To determine the total square yards of sod needed for a particular landscaping site

MATERIALS
sod area
calculator

PROCEDURE
The area to be sodded should be measured to determine how much sod is needed. Sod will give immediate coverage to the area, reducing erosion of the top soil and immediately improving the aesthetics of the area being sodded.

Certified sod is certified turfgrass sod that is a superior product grown from a mixture of approved certified seed. As this certified seed is grown in the field, it must be inspected throughout the year to ensure genetic purity; density, color, and texture; aesthetic appearance; and that it is free of noxious weeds, diseases, and insects. The State Department of Agriculture will issue certification labels that are attached to the bill of laden (delivery ticket).

Certified sod is sold by the square yard. To determine the square yards of sod, multiply the length in feet by the width in feet and divide by 9 square feet per yard.

Formula: $\dfrac{\text{length in feet} \times \text{width in feet}}{9 \text{ square feet per yard}} = \text{square yards}$

NOTE: remember there are 43,560 square feet per acre.

1. Solve the following problems.

	Length in Feet	× Width in Feet	=	Square Yards
1.	250′	250′		_____
2.	210′	205′		_____
3.	5′	375′		_____
4.	15′	625′		_____
5.	106′	125′		_____
6.	560′	356′		_____
7.	32′	90′		_____
8.	3′	65′		_____
9.	12′	725′		_____
10.	45′	37′		_____
11.	73.5′	62.5′		_____

	Length in Feet	× Width in Feet	=	Square Yards
12.	92.5′	107.5′		_____
13.	75.75′	85.25′		_____
14.	129.5′	185.5′		_____
15.	7.5′	195.5′		_____

2. If the cost of sod is \$1.35 per yard, what would be the cost of sod in 1 through 15 above?

1. _____

2. _____

3. _____

4. _____

5. _____

6. _____

7. _____

8. _____

9. _____

10. _____

11. _____

12. _____

13. _____

14. _____

15. _____

LAB EXERCISE 8–3

Lawn Seeding

PURPOSE

To practice the establishment of a lawn by seeding

MATERIALS

area of 1000 sq. ft.
turf seed
rototiller
cyclone-spreader
lawn roller
soil test kit
lawn tools: steel rake, flat shovel, wheelbarrow, round point shovel, lawn sprinkler, rain gauge, nursery spade
lime, sulfur, and fertilizer
straw mulch

PROCEDURE

1. Select an area of 1000 square feet.

2. Rototill the area.

3. Rake and level the area.

4. Take soil samples and test for the pH and N,P,K.

5. Apply lime or sulfur to correct the pH level and rake it in the soil. If the pH soil test is lower than 6.0, add lime to correct the pH level. If it is higher than 7.5, add sulfur to lower the pH level. Follow the manufacturer's recommendation on the bag.

6. Use the cyclone spreader to apply fertilizer according to the soil test and rake it in to the soil.

7. Using the cyclone spreader, seed the grass at the rate recommended on the bag.

8. Seed one half of the seed in a north/south direction and one half in east/west direction.

9. Using the steel rake, rake in the seed and lightly cover it with the soil.

10. Use the lawn roller to firm the seed into the soil. (Note: the rolling will cause capillary action movement of the soil water which will promote seed germination.)

11. Apply straw mulch using one bale (60 lb.) to cover 1000 square feet.

12. Set up the lawn sprinkler to moisten the seed, soil, and straw. Apply at least 1 inch of water at each watering. (Set up a rain gauge to be sure you have applied an inch of water at each watering.)

13. Check daily for the results of germination. It will take about 8 to 10 days for germination of the seed.

14. Water as needed.

LAB EXERCISE 8–4

Plant Science
Side 2
Turfgrass Field Operations

Lawn Renovation (Re-seeding)

PURPOSE

To re-seed a presently established lawn that is thinning or fill in where dead spots have occurred

MATERIALS

Lesco lawn renovator No. 20
selected lawn seed for your area
soil test
lime or sulfur
fertilizer 5-10-5
lawn sprinkler and rain gauge
Introductory Horticulture, 5th Edition

PROCEDURE

1. Soil test the area to be renovated for pH and N,P,K.

2. Apply lime or sulfur to correct the pH to a steady 6.5.

3. Using a Lesco lawn renovator, overseed the lawn area.

4. Set up the sprinkler and rain gauge to apply 1 inch of water at each watering.

5. Observe for germination of the grass in about 8 to 10 days.

LAB EXERCISE 8–5

Lawn Renovation (Over Seeding)

PURPOSE

To overseed an established lawn or athletic field of thinning turf

MATERIALS

Lesco lawn aerator 30
Lesco hi-wheeled cyclone spreader
selected lawn seed for your geographical area
soil test kit
lime or sulfur (for pH correction)
lawn starter fertilizer
lawn sprinkler and rain gauge
chain linked fence section 4' x 5' for dragging
Introductory Horticulture, 5th Edition
Landscaping by Jack Ingels, Delmar Publishers

PROCEDURE

1. Soil test the area to be overseeded for pH and N,P,K.

2. Correct soil pH and N,P,K to the soil test recommendation.

3. Using the Lesco Aerator 30, aerate north/south and east/west. This will ensure good penetration.

4. Using Lesco hi-wheeled spreader, sow the grass seed over the area, using local recommendations on seed and seeding rate.

5. Using a 4' × 5' section of a chain linked fence, drag the seeded area to ensure good soil/seed contact.

6. Set up the sprinkler and rain gauge to apply 1 inch of water over the entire seeded area.

7. Observe for germination of the seed in 8-10 days.

LAB EXERCISE 8–6

Lawn Watering

PURPOSE
To learn how to water a lawn with a lawn sprinkler

MATERIALS
lawn sprinkler
garden hoses
nursery spade
ruler
soil auger or soil probe
2 rain gauges or several containers such as coffee cans or glass beakers

PROCEDURE
Less than 1 percent of all the water on the Earth is fresh. Water is the life blood of plants. It is important to know when plants need to be watered in the landscape. The amount of water applied will depend on two factors: the plant's requirements for water and the soil type. Plants will vary in their demand for water (e.g., hollies thrive in moist soils, while yews like well-drained soils and less moisture in the root zone). Other ecological factors will affect plant water demands too; but for this lab we consider plant requirements and soil type.

Soil type influences available water for the growth of plants. There are three types of water in the soil: hydroscopic, capillary, and gravitational. Gravitation water moves down through the soil, capillary water moves up through the soil, hydroscopic water is held in the soil by soil particles and is not available to the plants.

When watering plants, it is best to water thoroughly and infrequently. Water moves in the soil up, down, and laterally. This is influenced by the texture of the soil. The texture of the soil may be fine, medium, or coarse. These factors will have a direct influence on the available water. A benchmark to follow: 1 inch of water applied to an area will penetrate clay soils (fine texture) to 4 inches, loam soils (medium texture) to 8 inches, and sandy soils (coarse) to 12 inches.

1. Select a lawn area to be watered.

2. Use a lawn sprinkler to apply the water.

3. Place the two rain gauges or cans in the area to be sprinkled.

4. Turn the sprinkler on for 15 minutes and measure the amount of water in the rain gauges or cans. Repeat this process at 30, 45, and 60 minute intervals.

5. After each 15-minute measurement of the water in the rain gauge, record the measurement and probe the soil to determine the water penetration of soil. Record the depth of the moist soil.

Answer the following questions.

1. How does the soil texture influence the movement of water? _____

_____

2. Why is it important to apply at least 1 inch of water at each application?_____

_____

_____

3. What situation do you create when water is applied faster than it can soak into the soil surface?

_____

Why is this dangerous to the environment? _____

_____

_____

LAB EXERCISE 8–7

Calibrating a Cyclone Spreader

PURPOSE
To learn how to calibrate a cyclone spreader to apply the correct amount of materials

MATERIALS
10′ × 25′ sheet of black plastic
5-gallon bucket
scales to weigh the chemical
100-foot tape
Lesco hi-wheeled cyclone spreader
bag of pelleted limestone/or lawn fertilizer

PROCEDURE
The cyclone spreader is a major piece of landscape maintenance equipment in the horticultural industry. This equipment is used to apply lime, fertilizer, and pesticides. It important to know the correct amount of the compound being applied. When we apply more material than necessary it is detrimental to the environment.

1. Measure 20 linear feet in a selected area for trial applications for calibrating the spreader.

2. Weigh out 20 lb. of the material (e.g., pelleted limestone, fertilizer).

3. Close the spreader opening at the bottom by placing it in the off position.

4. Set the spreader opening at the rear of the spreader to C.

5. Fill the spreader with the 20 lbs. of material from step 2.

6. Spread the material over the pre-measured area. Please note how wide the cyclone spreader is throwing the material; on the average it will be approximately 5 feet.

7. After spreading the pre-measured area, weigh the remaining material in the spreader. Record the amount used.

8. Multiply the pre-measured 20′ × 5′. 100 sq ft. would be the treated area.

9. Multiply the weight by 10. This will equal the total pounds applied per 1000 sq ft.

10. Record the weight.

11. Repeat the procedure for each change of setting on the spreader.

12. Develop a line graph by plotting the settings to the total pounds of material spread. Use the blank graph on the next page to record your data.

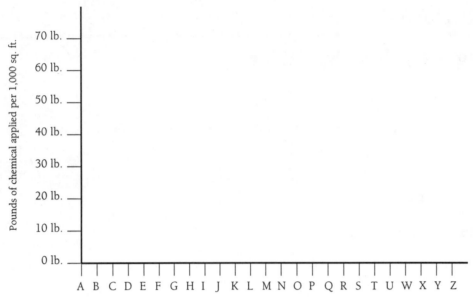

Cyclone Spreader settings

LAB EXERCISE 8-8

Plant Science
Side 2
Turfgrass

Operating a Commercial Walk-Behind Rotary Mower

PURPOSE

To learn how to operate a commercial walk-behind, rotary mower

MATERIALS

a commercial walk-behind rotary mower
lawn area for practice mowing
Lawn Mower Safety Video by ALCA

> **CAUTION**
> Use extreme care when handling
> Commerical Walk-Behind Rotary
> Mowers! Should be operated by
> trained individuals only!

PROCEDURE

1. Read the following information about a commercial rotary mower.

2. Instructor will demonstrate how to properly and safely operate the mower.

3. Answer the following questions about the material you have just read:

    A. When should spark spug wires be pulled? _____

    B. Should the engine be refueled when hot? _____

       Why or why not? _____

       _____

    C. Before using a mower, what should be checked prior to starting? _____

       _____

    D. How should slopes be mowed? _____

       _____

4. Demonstrate to your instructor your ability to operate the mower. (Your instructor will provide a lawn area to mow and will observe your use of the mower.)

5. Demonstrate to your instructor your ability to do daily servicing on the mower by completing the following tasks:

    A. change the oil on the mower;

    B. refuel the mower; and

    C. adjust mowing height.

6. Demonstrate to your instructor the proper care and use of the grass catcher.

COMMERCIAL WALK-BEHIND ROTARY MOWER

SAFETY PRECAUTIONS
A. General:
1. Read this Operator's Manual before starting the mower. Study the controls and learn the proper sequence of operation.
2. Do not allow anyone to operate or maintain this machine who has not read the manual. Never permit children to operate this machine.
3. Always have your feet and hands clear of the cutter deck when starting the engine.
4. Do not remove any shields, guard, decals, or safety devices. If a shield, guard, decal, or safety device is damaged or does not function, repair or replace it before operating the mower.
5. Always wear safety glasses, long pants, and safety shoes when operating or maintaining this mower. Do not wear loose-fitting clothing.
6. Never run the engine indoors without adequate ventilation. Exhaust fumes are deadly.
7. To avoid serious burns, do not touch the engine or muffler while engine is running or until it has cooled after it has been shut off.

B. Related to Fuel:
1. Gasoline is highly flammable. Respect it.
2. Do not smoke or permit others to smoke while handing gasoline.
3. Always use approved containers for gasoline.
4. Always shut off the engine and permit it to cool before removing the cap of the fuel tank.
5. If the fuel container spout will not fit inside the fuel tank opening, use a funnel.
6. When filling the fuel tank, stop when the gasoline reaches one inch from the top. This space must be left for expansion. Do not overfill.
7. Wipe up any spilled gasoline.

C. When Mowing:
1. Keep adults, children, and pets away from the area to be mowed.
2. Never use this mower without the discharge chute installed and set in the down position.

3. Mow only in daylight or proper artificial light.
4. Always remove debris and other objects from the area to be mowed.
5. Watch for holes, sprinkler heads, and other hidden hazards.
6. Reduce speed when making sharp turns.
7. Always have proper footing on slopes and hill sides and never operate when conditions are slippery.
8. Always keep both hands on the handles. Always walk, never run.
9. Never engage the blade clutch when the engine is running unless you are on grass that you intend to mow.
10. Be careful when crossing gravel paths or roadways until the blades stop rotating.
11. Never leave the mower unattended without disengaging the blade clutch, shifting the transmission into neutral, placing the neutral latch levers in the neutral lock position, shutting off the engine, taking the key from the ignition switch, and closing the fuel shutoff valve.
12. Always park the mower and start the engine on a level surface with the transmission in neutral, the blade clutch disengaged, and the neutral latch levers in the neutral lock position.
13. Shut off the engine and wait for the blades to stop rotating before removing the grass catcher.
14. If you hit a solid object while mowing, disengage the blade clutch, shift the transmission into neutral, place the neutral latch levers in the neutral lock position, and stop the engine.

Disconnect the spark plug wire and inspect for damage. Repair any damage and make sure the blades are in good condition and the blade bolts are tight before restarting the engine.

15. Do not mow excessively steep slopes. Mow across the slope, not up and down the slope.
16. Never raise the mower deck while the blades are rotating.
17. Never walk or stand on the discharge side of a mower with the engine running. Disengage the blade clutch if another person approaches while you are operating a mower.
18. Always disconnect the spark plug wire to prevent the engine from accidentally starting before performing any maintenance on this mower.
19. Keep the mower and especially the engine clean and free of grease, grass, and leaves to reduce the chance of fire and to permit proper cooling.
20. The operator presence control levers located at each handle are designed for your safety. Do not try to defeat their operation. If the blade clutch is engaged or the transmission is in gear, releasing both handles will shut off the mower's engine.

OPERATING INSTRUCTIONS

A. Controls.

1. Ignition Switch:

Located in the center of the control panel between the handles. When the key is inserted and turned clockwise 90 degrees, the engine can be started if the transmission shift lever is in neutral and the blade clutch is disengaged.

2. Fuel Shutoff Valve:

Located under the fuel tank in the opening of the handle mount frame. The handle turns 90 degrees. When the handle is in a horizontal position, it will shut off the flow of fuel to the engine. When it is turned to a vertical position, it will open and allow fuel to flow to the engine. Any time the mower is being trailered or, if the mower will not be in use for 30 minutes or more, close the fuel shutoff valve to prevent flooding the engine.

3. Engine Throttle:

Located on the right side of the control panel between the handles. Moving the throttle lever from the front to the rear will increase the engine speed from slow to fast. To start the engine, set the throttle all the way to the rear in the "Choke" position. After the engine starts, move the throttle halfway between slow and fast.

4. Recoil Starter:

Located on the top of the engine. To start the engine, set the throttle on "Choke" and grasp the starter grip and pull slowly until the starter engages. Then pull the cord rapidly to overcome compression, prevent kickback, and start the engine. Repeat if necessary with the throttle pushed halfway back to the "Slow" position.

5. Transmission Shift Lever:

Located under the control panel. The lever has six positions moving from the left to the right: Reverse Gear, Neutral, First Gear, Second Gear, Third Gear, and Fourth Gear. The lever must be in the Neutral position in order to start the engine. Never attempt to mow in Fourth Gear. Fourth Gear should be used for transport only.

6. Blade Clutch:

Located on the left handle just below the control panel. This is an over-center belt clutch and when the handle is pushed forward until it snaps to rest, it forces the idler pulley into the blade drive belt which causes the blade to rotate. When the handle is pulled back, the pressure on the belt is relieved and the blades will stop rotating.

7. Steering/Brake Levers:

There is a right-hand lever located beneath the outer end of the right handle and a left-hand lever located beneath the outer end of the left handle. Each lever operates independently and when squeezed against spring tension, lifts the idler pulley from applying pressure to the traction drive belt on the right or left side and applies the right or left side brake. When the levers are released, the mower will move ahead in a straight line if the engine is running and the transmission is engaged in a forward gear. Steering is accomplished by squeezing the steering/ brake lever on the side to which the turn is to be made.

8. Neutral Latch Levers:
Pivoted inside each handle there is a neutral latch lever which works with each of the steering/brake levers. When either of the steering/brake levers is squeezed and its neutral latch lever pushed forward and engaged in the neutral lock position, the steering/brake lever is held in a position where the idler pulley is not applying pressure to the traction drive belt and the brake is not quite engaged. The neutral latch levers should be engaged in the neutral lock position before starting the engine.

9. Operator Presence Levers:
Located above the outer ends of the right and left handles, these levers must be held down on the handles against spring pressure in order to shift the transmission shift lever into gear or engage the blade clutch. Releasing the operator presence levers with either the transmission shift lever in gear or the blade clutch engaged will shut off the engine.

B. Initial Adjustments
    1. Disconnect the spark plug wire.
    2. Check the tire pressure. Drive Wheels should be inflated to 25 psi. Caster Wheels should be inflated to 15 psi. Note: New tires are overinflated in order to properly seat the bead to the rim.
    3. Check that all nuts, bolts, and screws are tight.
    4. Check the tension of the deck drive belts:
       a. Remove the deck cover shield and engage the blade clutch.
       b. Make sure the belts clear the belt guides by ⅛″ to ¼″.
       c. The tension of the deck drive belts should be adjusted so that a ten-pound pull between two pulleys deflects each belt about ½″. Do not overtighten these belts. The blade clutch should engage with only moderate force.
       d. Replace the deck cover shield and disengage the blade clutch.
    5. The tension of the transmission drive belt should be adjusted so that a five-pound pull between the engine traction drive pulley and the transmission drive pulley deflects the belt about ³⁄₁₆″.
    6. The two drive wheel belts are self-adjusting.
    7. The steering control rods on each side of the handle assembly should initially be adjusted so that there is about a ¼″ space between the rod and the bottom of the slot in the neutral latch lever with the latch in the drive position. To make this adjustment, remove the large hairpin from the swivel joint at the bottom of each steering control rod and thread the swivel joint up or down the rod as needed.
    8. The brake rods (above each drive wheel) should be adjusted so that when the steering/brake levers are squeezed and the mower is pulled backward, the brakes lock the drive wheels. The machine should roll freely when the neutral latch lever is in the neutral lock position. To adjust the brake rods, remove the large hairpin from the swivel joint at the top of each brake rod and thread the swivel joint up or down the rod as needed.
    9. Adjusting the cutting height: The mower is shipped with the cutting height set at 3 inches ±¼ inch depending on the air pressure in the tires.
    10. Lubricate all fittings listed in the maintenance section.

C. Break-In and Operation.
1. Make certain you thoroughly understand all of the safety precautions before you attempt to operate this machine.
2. Check the engine oil level. Fill to the proper level with 10W-40 engine oil rated for service SE or SF.
3. Move the mower outdoors. Check the engine gasoline level. When filling the tank, stop when the gasoline reaches one inch from the top. This space must be left for expansion. Use fresh, clean, unleaded, regular gasoline.
4. Move the mower to a "test area" where you can operate the mower for about half an hour without being disturbed.
5. To start the engine:
   a. Shift the transmission to neutral.
   b. Disengage the blade clutch.
   c. Place the neutral latch levers in the neutral lock position.
   d. Connect the spark plug wire.
   e. Open the fuel shutoff valve.
   f. Move the throttle lever to the "Choke" position.
   g. Put the key in the ignition switch and turn the switch on. (For electric-start engines: Turn the key to "Start". Do not hold the key in the start position for more than 10 seconds or you may damage the starter.)
   h. For recoil-start engines: Grasp the starter grip and pull slowly until the starter engages and then pull the cord rapidly to overcome compression, prevent kickback, and start the engine. Allow the cord to recoil slowly. Repeat if necessary.
   i. Set the throttle at 50 percent of full engine RPM and allow the engine to warm up. Then, adjust the throttle to 75 percent of full engine RPM.
6. After the engine has warmed up, shut off the ignition and check the operation of the safety switches. Make certain that the engine will not start unless the ignition switch is turned on, the transmission is in neutral, and the blade clutch is disengaged. If the engine will start with the transmission in any gear other than neutral, immediately shut off the engine and replace the neutral safety switch on top of the transmission. If the engine will start with the blade clutch engaged, immediately shut off the engine and adjust or replace, if necessary, the blade safety switch mounted on the engine deck. Start the engine and hold the left operator presence control lever down against the left handle and engage the blade clutch. Now take your hand off of the operator presence lever and the engine should die. If it does not, immediately shut off the engine and adjust or replace, if necessary, the operator presence switch under the control panel. Disengage the blade clutch.
7. Restart the engine.
8. Push the blade clutch lever forward until it engages and the cutter blades start rotating.
9. Shift the transmission into first gear. (It is suggested that you practice mowing in first gear.)
10. Squeeze both steering/brake levers with both hands and release the neutral latch levers from the neutral lock position.
11. Slowly release the steering/brake levers and the mower will move ahead in a straight line. To turn the mower, squeeze the steering/brake lever on the side to which you want to turn.
12. To stop the mower's forward motion, squeeze both steering/brake levers until the mower stops and place the neutral latch levers into the neutral lock position.
13. Before shifting into reverse gear, the mower's forward motion must be completely stopped. For maximum traction in reverse, the steering/brake levers should be pushed downward.

 WARNING
To avoid possible bodily injury and to prevent damage to the transmission, the mower must be completely stopped before attempting to shift from forward to reverse or reverse to forward.

14. Practice operating the mower as you gain confidence, shift the transmission from first to second and mow for a while and then shift from second to third. Mow for at least one-half hour and then return the mower to the shop.

15. To stop and shut off the mower, squeeze both steering/brake levers and place the neutral latch levers into the neutral lock position, disengage the blade clutch, shift the transmission into neutral, turn off the ignition to stop the engine, close the fuel shutoff valve and disconnect the spark plug wire.

16. Check that all nuts, bolts, and screws are still tight.

17. Check, and adjust if necessary, the tension of the deck drive and the transmission drive belt as described in items 4 and 5 of the Initial Adjustment section.

18. Readjust the steering control rods and the brake rods. They may require frequent adjustment until the belts and brake banks have properly seated. These adjustments are described in items 7 and 8 of the Initial Adjustment section.

19. After the first full day of mowing, all nuts, bolt, and screws should be rechecked for proper tightness and the belts should be rechecked for proper tension.

MAINTENANCE

A. General Maintenance:

1. If the mower must be tipped on its side for maintenance, first drain the fuel from the fuel tank and the oil from the engine's crankcase.

2. Be careful not to spill lubricant on the drive belts.

3. Do not tamper with the engine's governor settings. They are adjusted to provide the proper maximum engine speed.

4. If the mower is to be in storage for more than 30 days, drain the fuel tank, run the engine to drain the carburetor dry, change the oil, remove the spark plug, and put a teaspoonful of oil into the cylinder. Pull the starter cord slowly to crank the engine and distribute the oil then replace the spark plug.

B. Daily Maintenance after Mowing:

1. Park the mower outside the storage facility with the engine shut off.
2. Close the fuel shutoff valve.
3. Permit the mower to cool.
4. Disconnect the spark plug wire.
5. Wash the mower off with a water hose. Be sure to clean out grass clippings from under the cutter deck and also under the deck cover. Allow the mower to dry before storing.
6. Check that the blade mounting bolts are tight.
7. Check that the blades are sharp. NOTE: Never mow with dull blades.
8. Check the fuel level, the engine oil level and clean the cooling-air intake (the rotary screen).
9. Clean the air cleaner elements (foam and paper).
10. After the first 5 hours of use, change the engine oil. (Change the oil every 40 hours thereafter.)
11. Follow the lubrication chart on the following page.
12. Place the mower in locked storage to avoid tampering or use by an untrained operator.

C. Maintenance Every 40 Hours:

1. Change the engine oil and replace the oil filter. (Change the engine oil more frequently under severe operating conditions.)
2. Check that all nuts, bolts, and screws are tight.
3. Check the condition and tension of all belts.
4. Clean the spark plug and check the spark plug gap.

5. Follow the lubrication chart below.

6. Check condition and operation of all safety switches.

D. Lubrication Chart

NUMBER OF GREASING POSITIONS

32" & 36"	48"	Item	Description

DAILY LUBRICATION CHART

32" & 36"	48"	Item	Description
2	3	A	Cutter Blade Spindle Bearings

40 HOUR LUBRICATION CHART

32" & 36"	48"	Item	Description
2	2	B	Caster Wheel Bearing
2	2	C	Caster Wheel Pivot Shafts
1	2	D	Deck Idler Pulley Pivot Arms
2	2	E	Drive Wheel Bearings
2	2	F	Brake Lever Pivots
1	1	G	Blade Clutch Bellcrank Pivot
2	2	H	Transmission-Jackshaft Couplers

E. Engine Maintenance:

For detailed maintenance instructions for the engine on your mower, see the Engine Manual packaged with your mower.

F. Mower Maintenance:

1. TO CHANGE A BLADE:
   a. Remove the deck cover.
   b. Tip the mower back and block up the front of the deck.
   c. Place one wrench on the hex head bolt under the blade. Use a second wrench to remove the locknut on top of the spindle pulley.
   d. Remember the number of blade spacers both above and below the spindle.
   e. Remove the long (9½") blade bolt, the flat washer, the blade and the blade spacers.
   f. To replace the blade, reverse the above procedure. Be careful to replace the blade spacers correctly above and below the spindle.

2. TO SHARPEN A BLADE:
   a. To sharpen a blade, use LESCO's Rotary Blade Grinder, No. 050655, or clamp the blade in a vise and, using as flat mill file, carefully file the cutting surface on each end of the blade to a sharp edge.
   b. Blades with severe nicks or dents that cannot be removed by filing should be replaced.
   c. After sharpening, check the balance of the blade by placing the blade on the LESCO Blade Balancer fulcrum device, No. 050532.
   d. If the blade dips on one end, file stock off the cutting surface on that end. NOTE: If a blade cannot be easily balanced, replace it.

3. TO CHANGE THE BLADE DRIVE BELTS:
   a. Make sure the blade clutch is disengaged.
   b. Remove the deck cover.
   c. Remove the cap screw which serves as a belt guide and is mounted in the idler pulley arm.
   d. Slip the long blade drive belt off of the pulleys.
   e. (48" ONLY) Loosen the idler pull rod which holds the idler pulley tight against the short blade drive belt.
   f. Remove the short blade drive belt from the pulleys.
   g. Place a new short blade drive belt back on the pulleys and tighten the idler pull rod to hold the idler pulley tight against the belt.

    h. Place a new long blade drive belt through the belt guide and loop it around the engine pulley and then around the two deck pulleys. The belt's back side should ride on the idler pulley.

    i. Replace the cap screw, lock washer, and nut in the idler pulley arm.

    j. The idler pulleys should be adjusted so that when the blade clutch is engaged, a ten-pound pull between two pulleys deflects either belt about ½″. Do not overtighten these belts. The blade clutch should engage with only moderate force.

    k. Replace the deck cover.

4. TO CHANGE THE TRANSMISSION DRIVE BELT:

    a. Make sure the blade clutch is disengaged.

    b. Working under the engine deck, take the long blade drive belt off of the engine pulley.

    c. Loosen the locknut holding the transmission drive belt idler pulley in place and slide the pulley away from the transmission drive belt.

    d. Remove the old belt and mount a new belt on the pulleys.

    e. Slide the idler pulley back onto the belt and tighten the locknut holding it in place. The idler pulley should be adjusted so that a five-pound pull on the belt between the engine pulley and the transmission pulley deflects the belt about $\frac{3}{16}$″.

    f. Replace the long blade drive belt on the engine pulley.

5. TO CHANGE EITHER TRACTION DRIVE BELT:

    a. Remove the hex nuts that hold the belt guard in place. Remove the belt guard.

    b. Place the neutral latch lever in the neutral lock position.

    c. Remove the hairpin cotters from the two swivel joints which are inserted through the idler bracket and remove the swivel joints from the idler bracket.

    d. Lift up on the idler bracket with one hand while removing the belt from the jackshaft pulley with the other hand.

    e. Prop up the side of the mower and slip the belt over the drive wheel.

    f. Slip the new traction drive belt over the drive wheel and loop it over the jackshaft pulley while lifting up on the idler bracket.

    g. Insert the swivel joints in the holes in the idler bracket and replace the hairpin cotters.

    h. Replace the belt guard and the two nuts.

6. TO CHANGE A SPINDLE ASSEMBLY:

    a. Make sure the blade clutch is disengaged.

    b. Remove the deck cover.

    c. Remove the blade (See paragraph 1. To Change a Blade.)

    d. Remove the blade drive belts. (See paragraph 3. To Change the Blade Drive Belts.)

    e. Remove the two bolts that hold the pulley bushing tight in the spindle pulley and thread the bolts into the other two holes in the bushing.

    f. Alternately turn each bolt ½ turn clockwise until the bolts force the pulley off the bushing.

    g. Remove the bushing by tapping a large screwdriver into the slot in the side of the bushing.

    h. Remove the key and the pulley.

    i. Tip the mower back and block up the front of the deck.

    j. Remove the four bolts and locknuts holding the spindle assembly to the deck.

    k. Remove the spindle assembly.

7. TO CHANGE SPINDLE BEARINGS:

    a. Clamp the spindle assembly in a vise with the grease fitting pointing up.

    b. Bend down the tab of the tab lockwasher.

    c. Remove the hex jam nut.

    d. Obtain a round wooden dowel about 6 inches long, such as a piece of broom handle. Hold this wooden dowel on top of the spindle and tap it with a hammer to drive the spindle out of the bearings and the spindle housing.

e. Pull the seal spacer out of the top seal.

f. Using a large screwdriver, pry the seals out of each end of the spindle housing. (NOTE: You will destroy the old seals as you remove them.) As you remove the old seals, the inner races of the roller bearings and the internal spacer will fall out of the spindle housing.

g. Using a bearing puller, remove the outer races from both ends of the bearing housing. NOTE: If the outer races cannot be removed using this procedure, take the spindle housing to a local machine shop and let them remove the bearing races.

h. Clean the old grease out of the spindle housing.

i. Clamp the housing in a vise with the grease fitting pointing down.

j. Gently drive a new bearing outer race back into the spindle housing using the wooden dowel. Make certain that the tapered side of the race faces outward.

k. Pack one inner race with fresh grease and place it into the outer race.

l. Press one of the new seals into the spindle housing by using the wooden dowel, moving around the circumference of the seal and tapping gently with a hammer. NOTE: The metal surface on the seal should face outward.

m. Turn the housing over and re-clamp it in the vise with the grease fitting pointing up.

n. Gently drive the second bearing outer race back into the spindle housing using the wooden dowel. Make certain that the tapered side of the race faces outward.

o. Insert the internal spacer.

p. Pack the second inner race with fresh grease and place it into the outer race.

q. Press the second new seal into the spindle housing by using the wooden dowel, moving around the circumference of the seal and tapping gently with a hammer. NOTE: The metal surface of the seal should face outward.

r. Coat the spindle with grease and push it up through the bottom of the spindle housing. The spindle will pass through the lower seal, an inner race, the internal spacer, an inner race, and the upper seal.

s. Hold the spindle in place and install the seal spacer down over the spindle and push it into the top seal. There is a notch on one of the inside edges of the seal spacer that should be facing up and should be opposite the side of the spindle with the keyway. Install the new tab lockwasher with its bent inner tab pointing down to engage with the notch in the seal spacer. Install the hex jam nut and tighten it. Bend up the large tab on the tab lockwasher to hold the jam nut in place.

LAB EXERCISE 8–9

Sharpening a Mower Blade

PURPOSE

To learn how to sharpen a mower blade. A sharp blade gives a clean cut of the grass.

MATERIALS

rotary mower blade
blade balancer
portable grinder
safety glasses
bench vise
wire brush

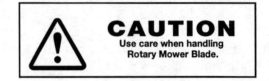

PROCEDURE

1. Secure the dull mower blade into a bench vise for grinding.

2. Clean the blade with a wire brush.

3. Sharpen the mower blade with a portable grinder to the manufacturer's specification angle. Wear your safety glasses.

4. Use the blade balancer to be sure equal amounts of the metal are removed from each side of the blade.

5. Sharpen the blade to the correct manufacturer's angle.

LAB EXERCISE 8-10

Operating a Commercial Dethatcher

PURPOSE

To operate a commercial dethatcher

MATERIALS

Lesco Commercial Dethatcher
lawn practice area

PROCEDURE

1. Read the following information

2. Instructor will demonstrate how to properly and safely operate a Lesco Commercial Dethatcher

OPERATING A COMMERCIAL DETHATCHER

SAFETY PRECAUTIONS

1. Read this manual thoroughly before operating or maintaining the dethatcher.

2. Always wear safety glasses, long pants, and safety shoes when operating or maintaining this machine. Do not wear loose-fitting clothing.

3. Always disconnect the spark plug wire before adjusting or maintaining this machine.

4. Never put your hands or feet under the deck assembly, except for maintenance with the spark plug wire disconnected.

5. Do not remove any shields, guards, or decals. If a shield, guard, or decal is damaged or does not function, repair or replace it before operating the dethatcher.

6. Altering this machine in any way may cause injury to the operator or bystanders.

7. Make certain the dethatcher is in the transport position when starting the engine.

8. When moving the dethatcher, always place the unit in the transport position and push the unit from behind.

9. Keep people and pets away from the dethatcher when in use.

10. Remove all foreign objects in the path of the dethatcher.

11. Watch out for sidewalks, curbs, rocks, small tree stumps, roots, sprinkler heads, stakes, water shutoff plates, etc.

12. If a breakdown occurs, shut off the engine immediately, disconnect the spark plug wire, and do not reconnect until repairs are made.

13. Never run the engine indoors without adequate ventilation. Exhaust fumes are deadly.

14. Gasoline is extremely dangerous and flammable. Do not permit open flames or sparks near the engine at any time.

15. Keep the dethatcher and especially the engine and belt area clean and free of grease, grass, and leaves to reduce the chance of fire and to permit proper cooling.

OPERATING INSTRUCTIONS PRIOR TO EACH DAY'S USE

1. Disconnect the spark plug wire.
2. Position the dethatcher on a level surface and check the engine oil and gear reduction oil levels. Add fluids as necessary.
3. Check the belt tension.
4. Check the condition of the blades. Adjust the depth control cam if necessary.
5. Check and tighten all nuts and bolts as necessary.

TO START THE DETHATCHER

1. Pull the control handle to the rear to lift the deck assembly to the transport position and push the dethatcher to the area to be dethatched.
2. Connect the spark wire.
3. Move the engine choke from "RUN" to "CHOKE". Move the engine throttle from "OFF" to "IDLE".
4. Grasp the starter grip and pull the cord slowly until the starter engages and then pull rapidly to overcome compression, prevent kickback, and start the engine. Allow the cord to recoil slowly. Repeat if necessary.
5. Move the choke back to "RUN" and after the engine has warmed up, move the throttle to "FAST".
6. Stand behind the dethatcher, move the machine to the point where you want to begin dethatching and push the control handle forward to lower the deck assembly so that the blades are in the dethatching position. Move slowly forward and begin dethatching. To make a turn, push down on the upper handle and pivot the dethatcher on its rear wheels with the blades disengaged from the turf. This will help to prevent bending or breaking the blades.
7. To shut off the machine, pull the control handle to the rear to lift the deck assembly to the transport position, move the engine throttle to "OFF" and disconnect the spark plug wire.
8. After use, thoroughly clean the machine, especially the engine, the belt area, and under the cutter deck.

LAB EXERCISE 8-11

Operating a Walk-Behind Blower

PURPOSE

To operate a walk-behind blower

MATERIALS

Lesco Walk-Behind Blower
lawn practice area

PROCEDURE

1. Read the following information
2. Instructor will demonstrate how to properly and safely operate a Lesco Walk-Behind Blower.

OPERATING A WALK-BEHIND BLOWER

SAFETY PRECAUTIONS

1. Read this manual thoroughly before operating or maintaining the blower.
2. Always wear safety glasses and long pants when operating or maintaining this machine. Do not wear loose-fitting clothing.
3. Always disconnect the spark plug wire before adjusting or maintaining this machine.
4. When operating, keep your hands and feet away from the side discharge and your hands and clothing away from the fan guard.
5. Do not remove any shields, guards, or decals. If a shield, guard, or decal is damaged or does not function, repair or replace it before operating the blower.
6. Altering this machine in any way may cause injury to the operator or bystanders.
7. Keep people and pets away from the blower when in use.
8. When operating, never direct the air stream at walls or vertical objects which can deflect debris back at you.
9. When operating, never direct the air stream at painted objects.
10. If a breakdown occurs, shut off the engine immediately, disconnect the spark plug wire, and do not connect until repairs have been made.
11. Never run the engine indoors without adequate ventilation. Exhaust fumes are deadly.
12. Stop the engine and allow it to cool before refilling the gas tank.
13. Gasoline is extremely dangerous and flammable. Do not permit open flames or sparks near the engine at any time.
14. Keep the blower and especially the engine air screen and cooling fins clean and free of grease, grass, and leaves to reduce the chance of fire and to permit proper cooling.

OPERATING INSTRUCTIONS

Prior to Each Day's Use:

1. Disconnect the engine's spark plug wire.
2. Position the blower on a level surface and check the engine oil level. Clean the cooling air intake screen. Replenish the fuel.
3. Check and tighten all nuts and bolts as necessary.

To Start the Blower:

1. Connect the engine's spark plug wire.
2. Move the engine choke control down to "CHOKE". Move the throttle control about 30 degrees counterclockwise.
3. Grasp the starter grip and pull the cord slowly until the starter engages and then pull rapidly to overcome compression, prevent kickback, and start the engine. Allow the cord to recoil slowly. Repeat if necessary.
4. Move the choke control up to the run position after starting and when the engine has warmed up, adjust the throttle to produce the amount of sweeping power required. At maximum throttle, the engine reaches the maximum speed of 3600 rpm which produces a fan tip velocity of 175 mph and the maximum blowing capacity of 2500 cfm.
5. Stand behind the blower and move to the point where you want to begin sweeping. The sweeping action is equally effective whether the machine is pulled or pushed but keep the front of the machine down and close to the work surface to blow out holes and crevices.
6. To shut off the machine, move the engine throttle control clockwise to the off position.
7. After use, thoroughly clean the machine, especially the engine's air intake screen and cooling fins.

ENGINE MAINTENANCE

Every 25 Hours: Service the foam air precleaner. Wash in soap and water, rinse and dry. Re-oil with clean engine oil. Change the oil. Change the oil more frequently under dusty conditions.

Every 50 Hours: Service the fuel filter.

Every 100 Hours: Clean the air cleaner paper element by gently tapping on the flat side. Do not wash or use pressurized air. Replace the paper element each year or more often under dusty conditions.

LAB EXERCISE 8-12

Operating an Aerator-30

PURPOSE

To how to operate a aerator to improve the aeration of the lawn

MATERIAL

Lesco Aerator-30
Lawn area to practice

PROCEDURE

1. Read following the information
2. Instructor will demonstrate how to properly and safely operate the Lesco Aerator-30

OPERATING AN AERATOR-30

SAFETY INSTRUCTIONS

1. The operator should have access to this manual at all times.

2. Do not operate this machine until you have read this entire manual thoroughly.

3. Make sure all guards are in place before starting this machine.

4. Never put hands or feet under the machine housing, except for maintenance. Always use handles when lifting.

5. Make certain the belt tensioner is not engaged when starting the engine.

6. Never stand in front of the machine while the engine is running.

7. CAUTION: The machine may start suddenly when the belt tensioner is engaged. Start machine at low throttle and gradually increase to desired operating speed.

8. If a breakdown occurs, shut off the machine immediately and do not restart until repairs are made.

9. Do not adjust the machine or perform any maintenance functions while the engine is running.

10. Never run the engine in an area without proper ventilation.

11. Gasoline is extremely dangerous and flammable. Do not permit open flames or sparks near the engine at any time.

12. Remove all foreign objects in the path of the aerator.

13. Do not run the aerator under low-hanging limbs which will interfere with the operator.

14. When transporting the unit, always push the unit from behind.

15. When parking the aerator on a slope, shut off engine, lower tines, and engage the belt tensioner. Never leave the aerator on a slope in the transport mode.

16. CAUTION: Do not exceed manufacturer's recommended tire pressure. Failure to comply may cause tire to explode.

17. Altering this machine in any way may cause injury to the operator or bystanders.

OPERATING INSTRUCTIONS

Prior To Each Day's Use:

1. Check engine oil and gear reduction oil levels.

2. Lubricate all grease fittings (4 wheels, 2 caster wheel yokes, spoon disc and shaft assembly, and jackshaft bracket).

3. Check belt and chain tension.

4. Check and tighten all nuts and bolts as necessary.

To Start The Aerator:

1. Disengage the belt tensioner.

2. Turn the engine switch to "on" and start the engine.

3. Adjust the throttle to a low rpm, firmly grip the handle and engage the belt tensioner.

4. To shut off the machine, disengage the belt tensioner and turn the engine switch to "off."

SECTION 9

The Vegetable Garden

◆

LAB EXERCISE 9–1

Establishing a Vegetable Garden

PURPOSE

To lay out a vegetable garden and produce vegetables for sale and for your home use

MATERIALS

graph paper
scale/ruler
pen or pencil
Introduction to Horticulture, 5th Edition
vegetable seeds and/or plants
gardening tools

PROCEDURE

1. Review Chapter 38 in your text.

2. Lay out two vegetable gardens: one for spring and one for late summer.

3. Using graph paper lay out two 20′ × 53′ gardens.

4. Select plants from those listed in the charts on the following pages. The key factors you must keep in mind while developing the garden layout: your personal preference invegetables; hardiness; spacing; days to harvest; quantity to grow; and days to germination.

5. In this layout, give the vegetable names and their required spacing.

6. Develop a vegetable garden at home as a part of your FFA-Supervised Agricultural Experience program.

7. Sell the vegetables as part of an agricultural entrepreneurship program.

NOTE: Plant selection will vary depending on geographic area and climate.

VEGETABLES AND TYPES	HARDINESS	DAYS TO GERMINATE	GROWING SUGGESTIONS	QUANTITY TO GROW	DAYS TO HARVEST	SATISFACTORY pH RANGE	USES
BEAN, snap bush and pole green, yellow	T	7–14	Sow bush types every 2 weeks until midsummer. Support pole types.	50 ft. bush; 8 hills pole	50–70	5.5–6.7	Fresh, frozen, canned. Vitamins A, B, C.
BEAN, bush shell red, white, green	T	7–14	Fava or English Broad Bean hardier than other types. Sow as early in spring as soil can be worked.	50 ft.	65–103	5.5–6.7	Fresh shell beans, or use dried for baking, soup, or Spanish or Mexican dishes. Vitamins A, B, C.
BEAN, lima bush and pole	T	7–14	Wait until ground is thoroughly warm before planting. Bush types mature earlier. Support pole varieties.	70 ft. bush; 8 hills pole	65–92	6.0–6.7	Fresh, frozen, canned, dried for baking. Vitamins A, B, C.
BEETS, red, golden, white	HH	10–21	For continuous harvest, make successive sowings until early summer. Do not transplant; this may cause forked or split roots.	25 ft.	55–80	6.0–7.5	Fresh, pickled, canned. Cook "thinnings" first, and tops later on for delicious greens. Vitamins A, B, C.
BROCCOLI	H	10–21	Plant again in midsummer for fall harvest. Grows best in cool weather.	25–40 plants	60–85*	5.5–6.7	Fresh, frozen. Vitamins A, B, C.
BRUSSELS SPROUTS	H	10–21	Pick lowest "sprouts" on stem each time; break off accompanying leaves but do not remove foliage.	25–40 plants	80–90*	5.5–6.7	Light frost improves flavor. Sprouts delicious fresh or frozen. Vitamins A, B, C.
CABBAGE, early, late, red, green	HH	10–21	Do not plant where any of the cabbage family grew the previous year.	25–40 plants	60–110*	5.5–6.7	Fresh, salads, coleslaw, sauerkraut. Winter storage. Vitamins A, B, C.
CARROT, long, short	HH	7–14	Short root types best for shallow or heavy soil. Plant again in midsummer for fall harvest.	25–30 ft.	65–75	5.2–6.7	Salads, relish, juice. Stews, soup. Vitamins A, B, C.
CAULIFLOWER, white, purple	HH	10–21	Tie leaves over heads to whiten.	16–24 plants	50–85*	6.0–6.7	Fresh, frozen, salad, relish. Vitamins A, B, C.

Note: See key on the last page of this figure.

VEGETABLES AND TYPES	HARDINESS	DAYS TO GERMINATE	GROWING SUGGESTIONS	QUANTITY TO GROW	DAYS TO HARVEST	SATISFACTORY pH RANGE	USES
CELERY	HH	10–21	To whiten, mound soil up around mature stalks.	30–36	115–135*	5.5–6.7	Raw in salads and as relish. Cooked and creamed, soups. Vitamin A.
CHARD red, white, stalked	HH	7–14	Pick frequently to encourage fresh leaves. Stands summer heat.	20–30 ft.	60	6.0–6.7	Cook leaves for greens; midribs and stalks like asparagus. Vitamins A, B, C.
COLLARDS	H	7–14	Easily grown, nonheading, cabbagelike leaves.	20–30 ft.	80	5.5–6.7	Cook leaves for greens. Popular in southern states. Vitamins A, B, C.
CRESS, garden and water	H	3–14	Sow in garden every 2 weeks for continuous supply. Also grows well on sunny windowsill. Grow watercress in moist, shady spots or along a shallow stream.	20–30 ft.	10–50	6.0–7.0	Salads, sandwiches, garnish, seasoning. Vitamins A, B, C.
CUCUMBERS, slicing, pickling	T	7–14	Grow on fence to save space. Keep picking to encourage new fruit.	8–12 hills	53–65	5.5–6.7	Salad, relish, pickles. Vitamin A.
EGGPLANT	T	10–21	Needs warm temperature—70° to 75°F for good germination. Pick fruits when skin has high gloss.	8–12 plants	62–75*	5.5–6.7	Delicious fried, sauteed, or in casseroles. Vitamin A.
ENDIVE	H	7–14	Grows best in cool weather.	20–30 ft.	90	6.0–7.0	Salad, greens. Hearts can be cooked and served with cream cheese or grated cheese. Vitamins A, B, C.
KALE	H	14–21	Mature plants take cold fall and winter weather. Frost improves flavor.	25–30 ft.	55–65	5.5–7.0	Chop young leaves for salads and sandwiches. Cook for greens. Vitamins A, B, C.
KOHLRABI	HH	14–21	Grow for spring or fall crop; thrives in cool weather.	16–20 ft.	55–60	5.5–6.7	Fresh, frozen; cooked like turnips. Vitamins A, B, C.
LEEK	H	14–21	Whiten and improve flavor by mounding soil around mature plants.	25–40 ft.	130	5.5–6.7	Fresh in salads. Cooked in soups, stews, or creamed.

VEGETABLES AND TYPES	HARDINESS	DAYS TO GERMINATE	GROWING SUGGESTIONS	QUANTITY TO GROW	DAYS TO HARVEST	SATISFACTORY pH RANGE	USES
LETTUCE, leaf	H	7–14	Make successive sowings in spring and another in late summer. Keep seedbed moist to get good germination for a fall crop.	25–40 ft.	40–47	6.0–7.0	Salad, sandwiches, garnish. Vitamins A, B, C.
LETTUCE, head	H	7–14	Needs cool weather in spring or fall to head well.	25–30 ft.	65–90	6.0–7.0	Salad, sandwiches, garnish. Vitamins A, B, C.
MUSTARD GREENS, fringed, smooth leaves	H	7–14	Grows as fall, winter, and spring crop in mild winter areas; spring and fall in north.	25–30 ft.	35–40	5.5–6.5	Greens. Vitamins A, B, C.
MELONS, cantaloupe, crenshaw, casaba, honeydew, watermelon	T	7–14	Very sensitive to frost. Black plastic mulch speeds maturity. Needs warm sunny weather when ripening for good flavor.	12–20 hills	75–120	6.0–6.7	Fresh, frozen. Ripe cantaloupes slip easily from stems. Ripe watermelons sound dull and hollow when tapped. Vitamins A, B, C.
OKRA	T	7–14	Needs hot weather to mature well. Pick pods young.	16–20 ft.	52–56	6.0–7.0	Soups, stews. Vitamins A, B, C.
ONIONS, yellow, white	H	10–21	Grows best in fine, well-drained sandy loam soil.	50–100 ft.	95–120	5.5–6.7	Fresh, salads, pickled. Vitamins B, C.
PARSLEY, curled or plain leaves	H	14–28	Attractive edging for flower garden; pot herb on sunny windowsill in winter.	10–20 ft.	72–90	6.0–7.5	Salad, garnish, seasoning. Dries or freezes well. Vitamins A, B, C.
PEA, dwarf, tall	H	7–14	Plant as early as ground can be worked.	40–100 ft.	55–79	5.5–6.7	Fresh, frozen, canned, dried. Vitamins A, B, C.
PEPPER, sweet, hot	T	10–21	Needs warm temperature—70° to 80°F for good germination.	8–10 plants	60–77*	5.5–6.5	Salad, stuffed, relish, seasoning. Vitamins A, B, C.
PUMPKIN, large, small bush, vine	T	7–14	For huge "contest" pumpkins, let only 1 or 2 grow per plant.	12–20 hills	95–120	5.5–6.5	Fresh, canned, frozen. Vitamins A, B, C.

VEGETABLES AND TYPES	HARDINESS	DAYS TO GERMINATE	GROWING SUGGESTIONS	QUANTITY TO GROW	DAYS TO HARVEST	SATISFACTORY pH RANGE	USES
RADISH, red, white, black	H	7–14	Make successive sowings until early summer; again a month before fall frost.	15–30 ft.	22–60	5.2–6.7	Relish, salad. Vitamins B, C.
RUTABAGA	H	14–21	Grows best in cool weather.	20–30 ft.	90	5.2–6.7	Fresh. Winter storage. Vitamins B, C.
SPINACH, crinkled, smooth	H	7–14	New Zealand and Malabar take hot weather; other varieties cool.	20–40 ft.	42–70	6.0–6.7	Greens, frozen, canned. Vitamins A, B, C.
SQUASH, summer, bush, vine	T	7–14	Keep fruits picked so plants produce more.	8–12 hills	48–60	5.5–6.5	Fresh, frozen. Vitamin A.
SQUASH, winter, bush, vine	T	7–14	Black plastic mulch speeds maturity.	8–12 hills	80–120	5.5–6.5	Fresh, frozen, canned. Winter storage. Vitamin A.
SUNFLOWER	T	7–14	Use for screen plant. Protect maturing heads with bags to prevent bird damage.	25–50 ft.	80	6.0–7.5	Bird, poultry seed.
SWEET CORN, white, yellow	T	7–14	Plant in blocks of short rows for good pollination and well-filled ears.	50–100 ft.	63–90	5.2–6.7	Fresh, frozen, canned. Vitamins A, B, C.
TOMATO, red, pink, yellow	T	7–14	Hybrids especially need warm temperature—70° to 80°F for good germination.	16–20 plants	52–68*	5.2–6.7	Fresh, salad, canned, juice, pickled. Vitamins A, B, C.
TURNIPS, white, yellow	H	14–21	Grow best in cool weather.	20–30 ft.	35–60	5.2–6.7	Fresh, raw, or cooked. Leaves of some types for greens.

KEY
H — HARDY VARIETIES OF VEGETABLES AND FLOWERS—Tolerate cool weather and frost. Plant fall to early spring in zones G, H, I, J. Two to four weeks before last killing spring frost in all other zones.
HH — HALF HARDY VARIETIES OF VEGETABLES AND FLOWERS—Tolerate cool weather and very light frost. For earlier maturity, varieties that transplant well can be started inside or in a cold frame 6–10 weeks before last expected light frost.
T — TENDER VARIETIES OF VEGETABLES AND FLOWERS—Cannot stand frost; plant in spring after last frost date.
* — Time from when plants are set into garden.

LAB EXERCISE 9–2

Pizza Garden Project

PURPOSE
To create a patio pizza garden of herbs used in pizza

MATERIALS
1″ × 4″ × 8′ pine board
plaster laths (⅜″ × 1″ × 4′)
4d box coated common nails
½-inch dowel rod (carrier handle)
propagated herbs of thyme, rosemary, sweet basil, chives, savory, oregano, and marjoram
3-inch clay pots or 3½-inch square plastic pots

PROCEDURE

1. Build a wooden carrying box that is 7″ × 12″. Cut two pieces 1″ × 4″ × 7″ (this will be the two end pieces); cut 10 plaster laths 12″ (4 laths for the bottom, 4 laths for the sides, 2 laths for the handle braces); cut the dowel rod 11⅞″ for the handle.

2. Assemble the pizza garden carrier by nailing the laths to the bottom, then nail the sides in place. Using the two laths remaining, center one on each end of the pizza garden and nail in place. Assemble the handle for carrying your garden.

3. Pot six of your selected herbs for the pizza garden using 3-inch clay pots or 3½ inch square plastic pots. This garden will provide you with six tasty herbs to use to flavor your next pizza.

LAB EXERCISE 9–3

Determining Proper Amounts of Vegetable Seeds

PURPOSE

To determine how much seed is needed to produce the desired number of transplants

MATERIALS

pen or pencil

PROCEDURE

Using the "Vegetable Planting Guide" on the next page, determine the correct ounces of seed or transplants needed in each of the following situations.

1. How many ounces of seed are needed to produce 10,000 cabbage plants? _____

2. How many ounces of tomato seeds are needed to plant an acre of tomato plants? (Assume there are 16,000 plants per acre.) _____

3. How many ounces of cucumber seeds are needed to plant an acre of cucumbers? (Assume there are 5,000 plants per acre.) _____

4. What is the recommended spacing for tomatoes? _____

5. How many ounces of pepper seeds are needed to plant an acre of peppers? (Assume there are 15,000

   plants per acre.) _____

6. How many pounds of watermelon seed are needed to plant an acre? (Assume there are 10,000 plants per acre.) _____

7. Why is it a recommended practice to bring one bee hive per acre of honey bees on crops like cucumbers?_____

   _____

   _____

8. What is the best average temperature for the germination of vegetable crops? _____

9. What is meant by "the cabbage will bolt if the temperature is 35 to 50 degrees F for 10 or more days"? _____

   _____

10. What are the best soils for growing celery? _____

    _____

Vegetable Planting Guide*

Vegetable	Seeds per oz.	Transplants per oz.	Germination Temperature	Comments
Cabbage	7,500	5,000	70-80º	Bolting will occur if early planted crop is subjected to 10 or more continuous days of temperatures between 35-50º
Cauliflower	10,000	4,000	70-80º	Start in greenhouse 4-6 weeks before planting
Broccoli	10,000	5,000	70-80º	
Brussel Sprouts	8,500	5,000	70-80º	Rows 3 ft. apart; plants 15"
Head Lettuce	20,000	9,000	60-75º	Rows 2 ft. apart; plants 12-15"
Onions	9,500	5,000	65-80º	
Celery	70,000	15,000	60-70º	Muck soils or well drained medium-textures mineral soils with irrigation work best
Tomatoes	10,000	4,000	70-80º	Spacing: Determinate - row 4-5 ft. apart, plants 18-24" apart; Indeterminate - rows 5-6 ft. apart, plants 30-48"; Staked - rows 5 ft. apart, plants 18" apart (trimmed to double stem
Peppers	4,500	1,500	75-85º	Warm Season crop that makes its best growth at 70-75º
Eggplant	6,000	2,500	75-90º	Temperatures below 65º result in poor growth and fruit set
Cucumber	1,000	500	70-95	For earlier production and more concentrated yields, use gynoecious varieties
Muskmelon	1,000	600	76-95º	Hybrid's should be harvested no sooner than half slip. Full slip gives optimum fruit quality and storage.
Watermelon	300	175	70-95º	Sufficient numbers of pollinating insects must be present. Local bees are seldom adequate for melons, cucumbers, pumpkins. One hive per acre recommended.

*Wetzel Seed Company—Harrisonburg, Va.

SECTION 10

The Small Fruit Garden

◆

LAB EXERCISE 10–1

Planting Strawberries

PURPOSE

To produce strawberries of marketable quality in the home garden

MATERIALS

planting site
fertilizer
plant labels
string
mulch
cultivator or hoe
tape measure
shovel
measuring sticks cut to proper plant spacing length
fumigant
Introductory Horticulture, 5th Edition

PROCEDURE

1. Select a planting site. Cultivate the soil immediately before planting to kill germinating weed seeds. Be sure all clods are broken up and that the soil is firm. Fumigation of the soil or other chemicals may be used.

2. Lay out rows according to spacing requirements. Rows must be opened deep enough to plant the roots straight down.

3. Keep the plant roots moist before planting. Never expose roots to the drying sun or wind.

4. Set the plant in the row at the proper depth with the crown just at ground level as shown below. This is very important. Plants set too deep smother and die; plants set too shallow dry out.

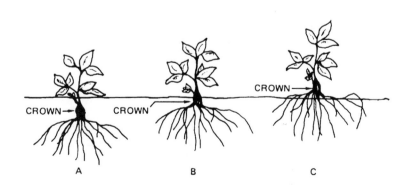

5. Fan the roots out and straighten them. If roots are too long, they may be trimmed to a length as short as 4 inches.

6. Pack the soil around the roots firmly. Do not cover the crown and do not leave any roots exposed.

7. If possible, water immediately after planting.

8. Enjoy your strawberries once they begin to grow!

Answer the following questions.

1. Draw and label the parts of a strawberry plant.

2. List three factors to consider when selecting varieties of strawberries for production.

_____

_____

_____

3. If you used the Hill System to plant strawberries with double rows:

   A. How many plants are needed to plant on an acre?   _____

   B. How many plants are needed to plant on an ½ acre? _____

   C. How many plants are needed to plant on an ¼ acre?   _____

   D. How many plants are needed to plant on an ⅒ acre?   _____

4. What method would you use in your garden for mulching strawberries?   _____

_____

_____

5. List the sixteen steps for better strawberry yields.

   _____   _____

   _____   _____

   _____   _____

   _____   _____

   _____   _____

   _____   _____

   _____   _____

   _____   _____

LAB EXERCISE 10–2

Blueberries

PURPOSE

To diagram a blueberry planting and select the best type of blueberry for your area

MATERIALS

pen or pencil
Introductory Horticulture, 5th Edition

PROCEDURE

1. List the six areas of the country in which blueberries grow naturally and the type of berry grown in each area.

Area	Berry Type

2. Identify the two types of commercially-grown blueberries.

_____

_____

3. Which type is best suited for growing in your local area? _____

4. What are the three main soil requirements for the best highbush blueberry production?

_____

_____

_____

5. Draw to scale a diagram of a blueberry planting.

Holiday Crafts and Floral Designs

◆

LAB EXERCISE 11–1

Designing a Della Robbia Wreath

PURPOSE
To design a Della Robbia wreath for home use

MATERIALS
wire box wreath frame
seed pods
cones
nuts
fruits
lotus pods
22-gauge florist wire
electric drill
hot glue gun
wire cutter

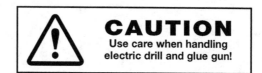

PROCEDURE

1. Select the desired size wreath frame for your project.

2. Soak any cones you are using in water for 30 minutes before the project is started. The water will cause the cones to close and they will be easier to work with when they are inserted in the wire wreath frame and while you are wiring them to the frame. As these dry out, the della robbia wreath will be much more condensed.

3. Place the frame face-down on a table. Insert the cones into the frame by pushing them in between the wires. Each cone is then wired in place with 22-gauge florist wire. The wire is attached to the cone and then woven through the frame and secured on the inside or back of the frame, concealing the wire from view. This process is continued until the outside and inside edges of the frame are lined with cones.

4. The center of the frame, which has been left open to this point, can be filled with cones placed upside down. This provides a contrast to the other two rows of cones. If a center row is formed with cones, clusters of seeds and nuts can be added to accent the wreath. The entire middle row can be filled with seed pods and nuts.

5. If nuts are being used as accents, it is best to drill a small hole through the nut (BE CAREFUL WHEN DRILLING. SECURE THE NUT; DO NOT HOLD IT IN YOUR HAND!) You can also use a hot glue gun which works well. BE CAREFUL OF THE HOT GLUE—IT WILL BURN YOU!

6. Other accents such as bows may be added for special occasions.

LAB EXERCISE 11–2

Constructing an Evergreen Wreath

PURPOSE

To construct evergreen wreaths for a class fundraiser at the holiday season

MATERIALS

common boxwood
box wreath frame
sphagnum moss
5 gallon bucket
22-gauge florist wire
plastic wreath wrap
Introductory Horticulture, 5th Edition
pruning shears
wire cutters

PROCEDURE

1. Select the desired size wreath frame for your project (a standard size for a door design is 16 inches). Review page 500 in your text.

2. Place the dry sphagnum moss in a 5-gallon bucket. Soak it with water until it is moist.

3. Place the moist sphagnum moss on the wire wreath frame evenly.

4. Holding the wreath frame firmly, wrap the packed sphagnum moss with the plastic wreath wrap. Be sure to pull the wreath wrap tight so that there are no gaps in the plastic. (Note: the tighter you make the plastic wrap the easier it will be to stick your evergreen boxwood in.) Then place a wire hanger on the back of the wreath. Fold the 22-gauge florist wire in half and loop through the wreath wire on the back side of the wreath form. Bend the florist wire to create the hanger for the wreath.

5. Select common boxwood sprigs which are about 4 to 6 inches long. The stems of common boxwood are stiff, which will make sticking the greenery sprigs easy and will puncture through the plastic wreath wrap. Place the individual sprigs of the boxwood in the wreath. Continue this operation in the same direction around the wreath form until the entire base is filled with the boxwood.

6. The completed boxwood wreath is attractive and has a colonial style. A bow can be added to accent the wreath.

LAB EXERCISE 11–3

Constructing a Hillman Wreath

PURPOSE
To construct a holiday wreath of marketable quality using the Hillman system

MATERIALS
HIllman wreath-making machine
Hillman wire wreath frames
evergreen material
pruning shears

PROCEDURE

1. Purchase a Hillman wreath-making machine and the wire wreath frames in the desired sizes.

2. Set up your Hillman wreath-making machine.

3. Collect the needed greenery: scotch pine, douglas fir, noble fir, balsam fir, Norway spruce, blue Colorado spruce, and other selected evergreens.

4. After you have collected the material, cut the greenery into 6 to 8 inch lengths, then gather them into bunches of 4 or 5 sprigs per bunch. To have the most attractive wreath, use the terminal bud shoots on top and cut ends under the terminal shoots. It will make a nice appearance on the finished product. On a 16-inch wreath form it takes about 20 to 24 bunches of 4 to 5 sprigs. Have these cut before you start the construction of the wreath.

5. To design the wreath, set the wreath form in the clamp shoe of the Hillman machine. Lay one bunch of the greenery on the wreath frame and clip it in place. Continue the construction of the wreath, keeping the greenery going in the same direction.

6. When you get to the last clip on the wreath frame, place the last bunch under the first bunch that you started with.

7. Be creative and mix the different greeneries—it makes for some interesting wreath designs.

8. Add bow for attractive accent.

LAB EXERCISE 11–4

Designing a Grapevine Wreath

PURPOSE

To design a grapevine wreath to be used during the holiday season

MATERIALS

grapevine—5 pieces, 4 feet long (grapevines may be collected from your vineyard after pruning or purchased from a florist wholesaler)
16-inch box wreath frame
pruning shears
wire cutters
20-gauge florist wire
hot glue gun
dried flowers or silk flowers
boxwood
florist knife

PROCEDURE

1. Using the box wreath frame as a form for the grapevine wreath, lay it flat on the workbench. Select the grapevines you will use. Wrap grapevine around the form, intertwining the grapevine to hold it in place. Once you have the grapevine wreath started, you can remove it from the form and continue to add more grapevines until you have the desired fullness of the wreath for your taste.

2. With the base completed, you may select dried flowers to create a crescent shape at the bottom of the wreath. Select a larger flower for the focal point and blend smaller flowers around that point and hot glue them into place. Use German statice as a filler flower and work this into the grapevine, following the crescent shape.

3. The 20-gauge florist wire may be used for creating the hanger for the grapevine wreath.

NAME: _____ DATE: _____

LAB EXERCISE 11–5

Designing a Holiday Topiary Arrangement

PURPOSE
To design a holiday topiary arrangement

MATERIALS
florist sheet moss
6-inch clay pot
1 foot white birch branch ½ inch to ¾ inch in
    diameter and 15 inches long
1 5-inch styrofoam ball
hot glue gun
3 lbs. of blue stone dust
10 to 15 hairpins (or use florist wire 3 inches long and bend it)
florist knife

PROCEDURE
1. Establish the foundation of the topiary design by placing the white birch branch in the center of the clay pot.

2. Now work the blue stone dust around the birch branch. Pack the dust tightly.

3. Continue this procedure until the blue stone dust is at the first rim on the clay pot.

4. Using the florist knife, cut a hole the same diameter of the branch into the styrofoam ball.

5. Using the hot glue gun, cover the area cut in the styrofoam ball with glue then quickly set the ball in place on top of the white birch branch and allow it to cool for 3 minutes.

6. Using the florist sheet moss, cover the styrofoam ball, pinning the sheet moss in place with hair pins. Cover the ball completely.

7. Cover the pot opening with sheet moss at the base of the topiary.

8. You may want to add small bows or other accessories depicting the holiday season.

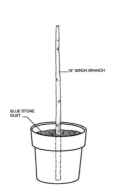

LAB EXERCISE 11–6

Creating a Holiday Centerpiece

PURPOSE
To create a holiday centerpiece

MATERIALS
6-inch floral container
white pine greens
variegated English holly
candle
toothpick or 18-gauge florist wire
Introductory Horticulture, 5th Edition
florist shears
florist knife

PROCEDURE
Review Unit 47 of your text and perform the following:

1. Prepare the base. Select a half block of floral foam and soak it in water until the foam is saturated. Place the foam in the 6-inch floral containers.

2. Insert the pine. Place the block on the design table. Take small clippings of white pine 4 to 5 inches long and push them into the foam around the base and carefully complete a line of greenery that will extend 4 to 5 inches out from the oasis. Design additional rows of greenery until the greenery reaches the top of floral foam.

3. To position the candle with variegated English holly, place the desired size candle with a tooth pick or 3-inch piece of 18-gauge florist wire in the base of the candle about 1 inch in, place the candle in the center of the floral foam block. Insert 4- to 5-inch sprigs of variegated English holly into the top portion of the floral foam. Use enough holly to cover the top of the floral foam block.

LAB EXERCISE 11–7

Creating a Miniature Christmas Tree Centerpiece

PURPOSE
To create a miniature Christmas tree centerpiece for the holiday season to use as a decoration at home

MATERIALS
1 block of floral foam
6-inch floral container
white pine greens, English boxwood, or
   variegated English holly
Introductory Horticulture, 5th Edition
florist shears

PROCEDURE
Review Unit 47 of your text and perform the following:

1. Prepare the base. Select a block of floral foam and soak it in water until the foam is saturated.

2. Shape the base by firmly placing the full block of oasis vertically into a standard 6-inch floral base. Using a piece of 18-gauge florist wire, shave the oasis to a tapered cone form by removing the corners of the oasis. Hold the florist wire firmly while making the cuts.

3. Prepare the greenery. Select your greenery from any number of narrowleaf or broadleaf evergreens. Prune the sprigs to 4 to 6 inches long.

4. Insert the greenery. Beginning at the base of the oasis cone, insert the sprigs evenly around the base, covering the container. As you move up the cone, you must maintain the natural taper and insert the greenery in its natural growing position. Also remember to shorten the sprigs of greenery to shape the mini-tree to taper with the intended design. Caution: when you get near the top, establish a center leader and use smaller greenery to avoid breaking the oasis on the taper.

5. Your finished tree should resemble the one pictured in your text.

6. Decorate the tree with selected ornaments, small bows, or cones in any desired color. The easiest way to fasten them in place is to use short pieces of florist wire and stick into the oasis.

7. To keep the arrangement fresh during the holiday season, water the arrangement by submerging the centerpiece in a tub of water enough to cover the entire arrangement. Allow this to soak until there are no longer any air bubbles coming out of the oasis.

LAB EXERCISE 11–8

Identifying Floral Ribbon

PURPOSE

To identify the different sizes of floral ribbon

MATERIALS

pen or pencil
tape
ribbon shears
belt of #40 ribbon
belt of #9 ribbon
belt of #5 ribbon
belt of #3 ribbon
belt of #1 ribbon

PROCEDURE

The bow is an important accessory for many floral design projects. The bow's design, texture, color, shape, proportion, balance, contrast, pattern style, and size will create the needed accent.

When buying the ribbon from a florist wholesaler, it is sold in full rolls called a bolt. The total length of the ribbon will vary with different manufacturers, but on the average there will be 25 yards on a bolt. Ribbon will be either single-faced or double-faced. The single-faced ribbon will have one shiny side and a dull texture on the other side, while the double-faced ribbon will have the same texture on both sides.

1. Using the ribbon shears, cut a sample of each of the ribbon and tape a sample of the different sizes next to the correct size on the next page. (See figure below.)

2. Using a scale or ruler measure the width of each ribbon size.

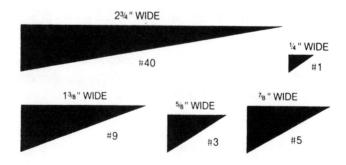

1. Size #40

2. Size #9

3. Size #5

4. Size #3

5. Size #1

LAB EXERCISE 11–9

Creating Ribbon Roses

PURPOSE

To learn how to construct ribbon roses to use as accessories in the construction of holiday arrangements

MATERIALS

single-faced #9 ribbon
22-gauge wire
floral tape
ribbon shears

PROCEDURE

Your instructor will demonstrate how to design a ribbon rose.

1. "A" in the following diagram indicates one end of the ribbon. Start the rose with the ribbon's shiny side facing up. Approximately 5 inches from this end, make a right angle fold as indicated. The working length of ribbon is called "B".

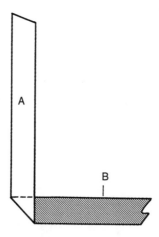

2. Hold the ribbon in your left hand with the "B" length straight out to the right. Take the "B" length with the right hand and fold it under and up, so it is parallel to "A". Make gentle folds, not sharp creases. The object is to form the ribbon into layers of squares.

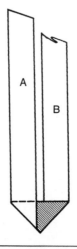

3. Hold the folded part in the left hand to keep it secure. Fold "B" away from you and down under to the left and on top of "A" (not behind) so as to form a square.

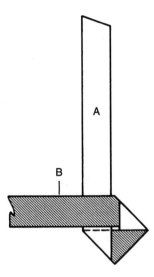

4. Hold the folded portion in the left or right hand, whichever is easier. Take the "B" length and fold it back and under at right angles bringing "B" straight down. To form the bow repeat this process.

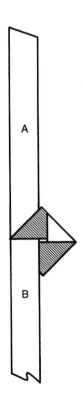

5. When three or four layers are completed, pull the "A" end through the center hole. Pull until about 3 inches come through.

Any ribbonend should be folded down and under at right angles. All of "B" is used up except for 2 inches.

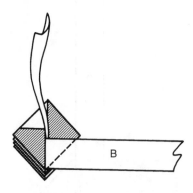

6. Turn the whole packet over so that the "A" end hangs down. Grasp the "A" end and twist slowly.

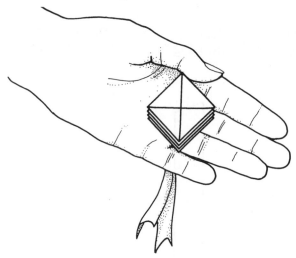

7. Take "A" and "B" ends and hold them tightly under the rose. Wrap 22-gauge florist wire around the ends and cover it with green florist tape.

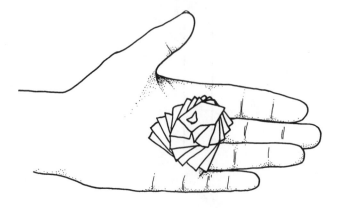

LAB EXERCISE 11–10

Creating a Circular Floral Arrangement

PURPOSE
To create a circular floral arrangement as a centerpiece on the dinner table

MATERIALS
baker's fern/leather leaf fern
baby's breath
6-inch floral container
½ block of floral foam
florist wire
wire cutter
florist knife
florist shears
Introductory Horticulture, 5th Edition

PROCEDURE
Review Unit 49 of your text and perform the following:

1. Select a 6-inch round florist container.

2. Soak the half block of floral foam in a bucket or water. Be sure it is saturated with water. Water is one of the key factors in the life of the arrangement. Place a half block of floral foam into the container. Firm the florist foam into place so the arrangement will be secure in the container after completing the design.

3. Select the baker's fern for greening of the arrangement. Arrange the fern to cover the entire block of floral foam and other mechanics.

4. Use the design rule of thumb of creating a design that is the design should be 1½ to 2 times the width of the container. Select the first carnation to establish the height of the arrangement by centering it with a 15° tilt (placing the flower straight up in the arrangement gives a falling-forward feeling while observing it).

5. Establish the circular outline at the base of the arrangement and maintain the same distance for this carnation as for the first one in step 4.

6. Turn the design so that the third carnation may be placed into the arrangement. This also allows the designer to see the floral arrangement from different angles.

7. Arrange the next carnations to establish the floral outline. With these points set, the basic foundation of the arrangement is organized.

8. Arrange two carnations in between the top and base of the design.

9. Add the remaining carnations to the arrangement until the circular shape of the arrangement is filled. It is important to maintain space between each carnation. This allows their natural beauty to radiate. The greenery should extend out around each carnation.

10. To complete the design, select filler flowers like baby's breath and arrange the filler flowers between the carnations. Allow the baby's breath to extend out beyond the carnations. It gives the arrangement more depth and movement.

LAB EXERCISE 11–11

Making a Carnation Corsage

PURPOSE

To make a single carnation corsage for prom night

MATERIALS

standard carnation
baker's fern/leatherleaf fern
florist shears
22-gauge florist wire
florist tape
florist stapler
#3 ribbon for the bow
corsage pin
corsage bag
florist knife

CAUTION
Use care when handling sharp
florist shears and knife.

PROCEDURE

1. Select a top standard carnation. The color is based on individual preference. Remove the stem half of an inch below the calyx.

2. Using the piecing method, push the florist wire through the calyx (see figure below) with an 8-inch piece of 20-22 gauge wire. Pull the wire straight down beside the stem; do not twist the wire. This will make the stem too bulky. Tape the flower by holding the flower in one hand, pull and twist as the tape is applied to the calyx, stem, and wire. Using the ½-inch florist tape, cover all exposed calyx, stem, and wire. Remember to keep the tape tight and pull it because it will be more adhesive with the twisting motion. Leave the stem about 4 inches long.

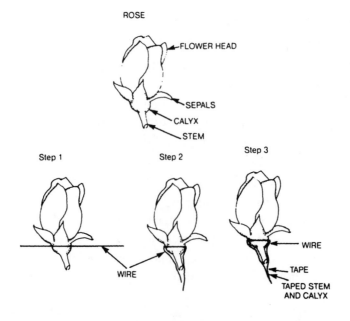

ROSE

FLOWER HEAD

SEPALS

CALYX

STEM

Step 1   Step 2   Step 3

WIRE

WIRE

WIRE

TAPE

TAPED STEM
AND CALYX

3. Select a piece of greenery for the backing of the corsage. Baker's fern will be used in this project. Tape it to the back of the carnation with ½-inch tape. Be sure it is in the proper proportion to the flower to ensure the correct artistic appeal.

4. Make a pinch bow (see below). In making the bow, be sure the bow is the same width as the flower when making the loops. Personal taste will also dictate how large the bow is made.

5. Take the taped carnation with the baker's fern and add the bow to the underside of the carnation. Tape the bow into place with the short piece of florist tape. Using the florist shears, remove about one inch of the taped stem.

6. Place a corsage pin in the back of the corsage by inserting it vertically into the stem. This way no one will be injured with the corsage pin.

7. Place the corsage into a corsage bag. Using a spray bottle moisten the inside of the bag. Fold the bag and staple.

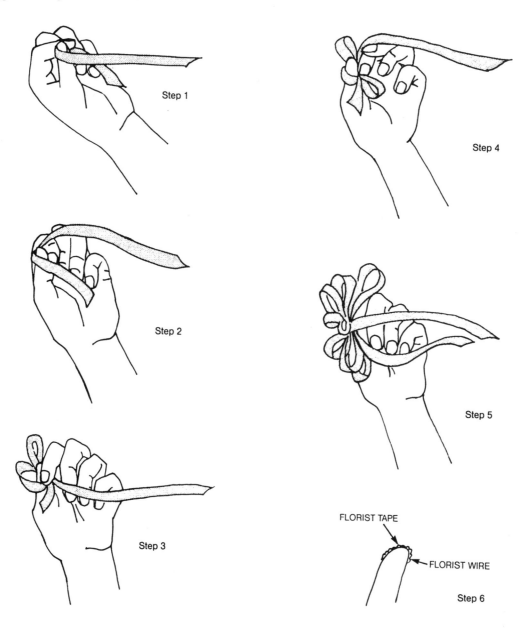

Step 1

Step 2

Step 3

Step 4

Step 5

FLORIST TAPE

FLORIST WIRE

Step 6

LAB EXERCISE 11–12

Making a Feathered Carnation Corsage

PURPOSE
To make a feathered carnation corsage for a flower girl in a wedding

MATERIALS
standard carnations
baker's fern/leatherleaf fern
florist shears
22-gauge florist wire
florist tape
florist stapler
#3 ribbon for the bow
corsage pin
corsage bag
florist knife

CAUTION
Use care when handling sharp
florists shears and knife.

PROCEDURE

1. To create this corsage, select two large carnations. To feather a carnation bloom, use a sharp florist knife and cut one of the blooms in half. Using 9-inch pieces of 24-gauge florist wire, wrap the remaining calyx of one of the flower halves. (NOTE: hold the flower firmly as the wire is twisted around the calyx.) Tape the carnation calyx and wire. Repeat the same process for other half of the flower.

2. Take the other standard carnation flower and with a sharp florist knife, split the calyx about half an inch on each side. This will allow the flower to open more widely. This flower will be used as the focal point in the corsage. Using the piecing method at right angles to the cuts in the calyx, wire and tape this flower.

3. For greenery in this arrangement, choose three short tips of carnation foliage. Using three pieces of 4-inch, 28-gauge florist wire, tape a foliage tip to each of the individual wires.

4. Lay out the three flowers as taped along with the three individual carnation foliage tips.

5. Design the corsage (see figure below).

6. Take one of the feathered carnations and one of the carnation foliage tips and press them together. The stickiness of the tape will hold them sufficiently until you can put them together with a short piece of florist tape.

7. Add the next feathered carnation, moving down the main stem of the corsage. Press this flower into place and tape it in the main stem.

8. Add the two remaining foliage tips below the last feathered carnation and on each side of the stem. Again, press the taped foliage tips to the stem and tape them into place with a short piece of florist tape.

9. Now add the focal point, the standard carnation, taping it at the base of the two feathered carnations.

10. Select a bow that will harmonize with the corsage and tape it below the last flower. Trim the long stems, but be sure to leave them straight to balance the corsage (see figure below).

LAB EXERCISE 11–13

Making a Rose Boutonniere

PURPOSE
To make a rose boutonniere for that special person to wear to the prom

MATERIALS
standard rose
baker's fern/leatherleaf fern
florist shears
22-gauge florist wire
florist tape
florist stapler
#3 ribbon for the bow
boutonniere pin
boutonniere bag
florist knife

CAUTION
Use care when handling sharp
florists shears and knife.

PROCEDURE
1. Select a high-quality flower.

2. Using the piercing method, attach a 6-inch piece of 20- to 22-gauge wire to the stem.

3. Bend the florist wire down next to the stem.

4. Select a piece of greenery to use on the back of the boutonniere. In this case, use baker's fern. Be sure to keep the greenery in proportion to the size of the rose, as it is used to complement the rose. (See figure on next page.)

5. Tape the baker's fern, rose stem, and wire all in one step. Be sure the tape is applied tightly and evenly down the stem. After taping, cut off the excess stem. The stem must be kept in proportion to the flower. Leave the stem straight because it gives a more natural appearance on the lapel.

6. Place a boutonniere pin in the back of the boutonniere by inserting it vertically into the stem. This way no one will be injured with the pin.

7. Mist the boutonniere.

8. Place the boutonniere into a boutonniere bag. Using a spray bottle moisten the inside of the bag. Place the boutonniere in a small bag or box. Fold the bag and staple it closed.

ROSE
Step 1

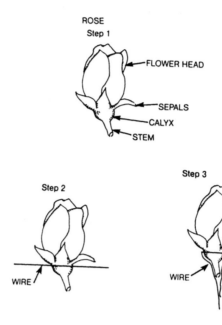

FLOWER HEAD

SEPALS

CALYX

STEM

Step 2

Step 3

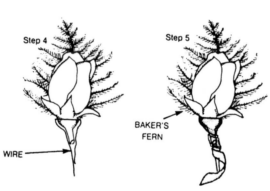

WIRE

WIRE

Step 4

Step 5

WIRE

BAKER'S
FERN